MICROBIOLOGY
LABORATORY MANUAL

PRINCIPLES
AND APPLICATIONS

MICROBIOLOGY
LABORATORY MANUAL

PRINCIPLES AND APPLICATIONS

Stephen A. Norrell
University of Alaska

Karen E. Messley
Rock Valley College

PRENTICE HALL
UpperSaddle River, New Jersey 07458

Acquisition Editor: *Linda Schreiber*
Executive Editor: *David Kendric Brake*
Editorial Director: *Tim Bozik*
Editor-in-Chief: *Paul F. Corey*
Assistant Vice President of Production and Manufacturing: *David W. Riccardi*
Editorial/Production Supervision: *Jennifer Fischer*
Executive Marketing Manager: *Kelly McDonald*
Marketing Assistant: *David Stack*
Total Concepts Manager: *Barbara Murray*
Manufacturing Buyer: *Ben Smith*
Manufacturing Manager: *Trudy Pisciotti*
Creative Director: *Paula Maylahn*
Art Director: *Heather Scott*
Cover Designer: *Paul Gourhan*
Cover Photo: Courtesy of *Thomas Broker/CNRI/Phototake*
Color Insert: Courtesy of *George A. Wistreich*

© 1997 by Prentice-Hall, Inc.
A Simon and Schuster/Viacom Company
Upper Saddle River, New Jersey 07458

Printed in the United States of America

10   9   8   7   6   5   4   3   2

ISBN 0-13-255373-2

Prentice-Hall International (UK) Limited, *London*
Prentice-Hall of Australia Pty. Limited, *Sydney*
Prentice-Hall Canada Inc., *Toronto*
Prentice-Hall Hispanoamericana, S.A., *Mexico*
Prentice-Hall of India Private Limited, *New Delhi*
Prentice-Hall of Japan, Inc., *Tokyo*
Simon & Schuster Asia Pte. Ltd., *Singapore*
Editora Prentice-Hall do Brasil, Ltda., *Rio de Janeiro*

CONTENTS

PREFACE

ABOUT THE MANUAL

This manual has been written for the Introductory Microbiology student enrolled in either a general or an allied health program. It is intended for those students who take microbiology during their freshman or sophomore year and who, typically, have had only one year (or less) of chemistry. Its aim is to familiarize the student with the basic procedures and equipment used in the microbiology laboratory—including diagnostic microbiology, clinical samples and sampling, serological procedures, antibiotic sensitivity testing and so on.

The manual has several important features. Each exercise has been written to be reasonably self-contained. It contains explanations of the salient points being demonstrated or tested. These explanations are meant to supplement and complement, not replace, the assigned text. The exercises have been prepared in a manner that will provide flexibility in their application, dependent on the time frame of the laboratory. This will also enable the exercise to be divided up for performance by individual members of the laboratory group.

DOING YOUR PART

Each exercise is designed to teach a new technique or illustrate a principle of microbial growth. They are arranged in the following sequence:

- Background material discussing the theory and experimental design underlying the procedures used in the exercise;

- Objectives of the laboratory exercise;

- Materials needed for the laboratory;

- Procedures outlining the specific instructions for completing the laboratory exercise; and

- Laboratory Report Form.

It is imperative that each exercise be read prior to attending the lab. Notice that a page for notes is included with each exercise. This page can be used to help you organize your laboratory exercise. It is often beneficial to write your own laboratory protocol checklist for the exercise. As each step is completed, it can be checked off, minimizing some of the confusion that occasionally occurs during the lab. Additionally, any changes made by your instructor can be noted here.

As you proceed through the laboratory manual, you may notice that the laboratory exercises get progressively more complex and the instructions get somewhat less comprehensive. You are expected to become more competent with each exercise and should be able to problem-solve some of the needed procedures for yourself.

Laboratory reports are required for all exercises, and report forms are included for each exercise. They require tabulation of the data acquired in the exercise and include questions based on the experimental procedures and principles demonstrated in the exercise. Many of the questions will require your application of the observed procedures or results to other situations.

LABORATORY SAFETY RULES AND REGULATIONS

Microbiology laboratory exercises are somewhat different from those you might have experienced in most other courses. Microbiology demands the use of live organisms. Most of the experiments in this manual require that you make observations and tests on living and actively growing bacteria. Some of the bacteria are opportunistic pathogens, however, and can be associated with laboratory infections. You must, therefore, use caution to prevent accidental infections. It is always wisest to assume that any culture you are using is pathogenic—then we are less likely to allow ourself to become careless.

There is also another important reason to develop good techniques for handling material that is potentially infectious. It is very likely that over the course of your career, you will be asked to handle infectious material. Syringes, urine samples, clothing—virtually any item that has been used by a patient is potentially infectious. As a health professional, you are obligated to do all you can to prevent spread of any diseases, including those encountered in your professional activity. There are also many instances where the maintenance of sterility is critical (e.g., surgery) and where the introduction of even noninfectious bacteria is a serious circumstance.

During this course, you will be introduced to the safe handling of materials containing microorganisms. With increased practice, you will be able to perform exercises so that the microorganism being tested will remain in the desired glassware and environmental microorganisms will be kept out. This is what is commonly referred to as aseptic technique. It

is essential for any work with microorganisms. Good laboratory safety practices include the following:

1. Laboratory exercises should be read before the laboratory period and work should be planned.

2. Upon entering the laboratory, place coats, books, and other paraphernalia in specified locations—never on bench tops.

3. Know the location of all safety equipment, including fire extinguishers and/or blankets and eye wash stations.

4. Wash your hands carefully with liquid soap before and after working in the laboratory. Dispose of the paper toweling in a safe way. In addition, hands should be washed immediately and thoroughly if contaminated with microorganisms.

5. Carefully wash your work area with disinfectant before you begin your work and after you complete it. This will limit the spread of dust-borne organisms as well as destroy any organisms that may have contaminated the work area.

6. Never eat, drink, smoke, or place anything in your mouth in the laboratory. Your instructor will give you special instructions for pipetting liquids containing bacteria.

7. Wear a lab coat or apron while working in the laboratory to protect clothing from contamina-

tion or accidental discoloration by staining solutions. In addition, wear shoes at all times in the laboratory, and tie long hair back.

8. Be especially careful to develop your aseptic techniques as they are presented to you. Most laboratory infections are the result of poor or lapsed technique.

9. Don't use equipment without instruction.

10. Horseplay will not be tolerated in the laboratory.

11. Do not place contaminated instruments, such as inoculating loops, needles, and pipettes, on bench tops. Loops and needles should be sterilized by incineration, and pipettes should be disposed of in designated receptacles.

12. Work only with your own body fluids and wastes in exercises requiring saliva, urine or blood to prevent transmission of disease.

13. Immediately cover spilled cultures or broken culture tubes with paper towels and then saturate them with disinfectant solution. After 15 minutes of contact, remove the towels and dispose of them in a manner indicated by your instructor.

14. Report accidental cuts or burns to the instructor immediately.

15. Do not touch broken glassware with your hands. Use a broom and dustpan to clean it up and place the broken glassware in a container as indicated by your instructor.

16. Inoculated media placed in the incubator must be properly labeled and put in the designated area.

17. On completion of the laboratory session, place all cultures and materials in the disposal area as designated by your instructor. All reagents and equipment used in lab must be returned to their proper place.

18. Never remove anything from the laboratory that has not been sterilized without the permission of your instructor. This is especially true for bacterial cultures.

LABORATORY SAFETY

RULES AND REGULATIONS

After you have carefully read the following rules and are sure that you understand them, sign this sheet and return it to your instructor.

1. Laboratory exercises should be read before the laboratory period and work should be planned.

2. Upon entering the laboratory, place coats, books, and other paraphernalia in specified locations—never on bench tops.

3. Know the location of all safety equipment, including fire extinguishers and/or blankets and eye wash stations.

4. Wash your hands carefully with liquid soap before and after working in the laboratory. Dispose of the paper toweling in a safe way. In addition, hands should be washed immediately and thoroughly if contaminated with microorganisms.

5. Carefully wash your work area with disinfectant before you begin your work and after you complete it. This will limit the spread of dustborne organisms as well as destroy any organisms that may have contaminated the work area.

6. Never eat, drink, smoke, or place anything in your mouth in the laboratory. Your instructor will give you special instructions for pipetting liquids containing bacteria.

7. Wear a lab coat or apron while working in the laboratory to protect clothing from contamination or accidental discoloration by staining solutions. In addition, wear shoes at all times in the laboratory, and tie long hair back.

8. Be especially careful to develop your aseptic techniques as they are presented to you. Most laboratory infections are the result of poor or lapsed technique.

9. Don't use equipment without instruction.

10. Horseplay will not be tolerated in the laboratory.

11. Do not place contaminated instruments, such as inoculating loops, needles, and pipettes, on bench tops. Loops and needles should be sterilized by incineration, and pipettes should be disposed of in designated receptacles.

12. Work only with your own body fluids and wastes in exercises requiring saliva, urine or blood to prevent transmission of disease.

13. Immediately cover spilled cultures or broken culture tubes with paper towels and then saturate them with disinfectant solution. After 15 minutes of contact, remove the towels and dispose of them in a manner indicated by your instructor.

14. Report accidental cuts or burns to the instructor immediately.

15. Do not touch broken glassware with your hands. Use a broom and dustpan to clean it up and place the broken glassware in a container as indicated by your instructor.

16. Inoculated media placed in the incubator must be properly labeled and put in the designated area.

17. On completion of the laboratory session, place all cultures and materials in the disposal area as designated by your instructor. All reagents and equipment used in lab must be returned to their proper place.

18. Never remove anything from the laboratory that has not been sterilized without the permission of your instructor. This is especially true for bacterial cultures.

I have read the above rules and regulations and understand their meaning.

_____

(signature)

MICROBIOLOGY
LABORATORY MANUAL

PRINCIPLES
AND APPLICATIONS

INTRODUCTION TO MICROSCOPY

One of the major tools of microbiology, ever since van Leeuwenhoek first observed his "anima-cules," has been the light microscope. In this exercise, we will review the compound light microscope, its construction, and its correct use.

BACKGROUND

The microscopes used by van Leeuwenhoek were *simple microscopes,* using only a single lens, which he would grind himself. These microscopes were only able to magnify a total of approximately 300x. Today's light microscopes are *compound microscopes* which use two or more lenses. The use of multiple lenses and better sources of illumination enable us to significantly improve both magnification and resolution. *Resolution* is the ability to distinguish clearly between two points or objects. Remember, just because an image is enlarged does not mean that we can see it more clearly!

A thorough understanding of the compound light microscope, in terms of its functioning and applications, is very important; the compound light microscope is central to all microbiology laboratories. For the purposes of this discussion, we will consider the microscope as consisting of two systems: the illumination system and the lens system.

THE ILLUMINATION SYSTEM

Proper illumination is essential if we are to view anything clearly with the microscope. van Leeuwenhoek was limited to the light of the sun or candles to see his specimen. These have been replaced with incandescent bulbs which provide readily controlled intensity and color. The focusing system for the light source includes the *iris diaphragm* and the *substage condenser.* The light from the incandescent bulb can be adjusted by opening and closing the iris diaphragm. The focusing system for the light source is the condenser. It is designed to deliver a cone of light, with dimensions specified. The position of the condenser (how far below the stage it is) determines how accurately the diameter of the cone of light matches the opening of the objective lens.

It is important to remember that, as a rule, one must increase the light as one increases the magnification. Even so, it is possible to have too much light illuminating the field of vision. When this occurs, resolution and contrast are lost, making slide preparations very difficult to observe and interpret correctly.

THE LENS SYSTEM

The dual lens system found in a compound microscope consists of the *objective lens,* located closer to the sample, and the *ocular lens,* located closer to your eye. Most microscopes used in the microbiology lab have three objective lenses located on a rotating nose piece. These are the low power (10x), high dry (43x) and oil immersion (97x) objectives. The ocular lens of most microscopes magnifies 10x. The total magnification of the microscope is determined by multiplying the magnification of the ocular and that of the specific objective being used.

objective magnification x ocular magnification
= total magnification

In other words, to determine the total magnification obtainable with the high dry objective, one would multiply the magnification of the ocular (10x) by that of the high dry objective (43x) giving a total magnification of approximately 430x.

The maximum usable magnification of a microscope is not limited by the achievable magnification of the lenses, but rather by the *resolving power* of the microscope. Resolving power is the ability to distinguish two points as separate and distinct. *Magnification,* on the other hand, simply measures how many times the lens systems of the microscope increase the apparent size. To understand the difference between resolving power and magnification, you should consider what would happen if you used grainy film to make a highly enlarged photograph—the result would be a large, blurry picture. A microscope with high magnification but poor resolving power produces a large blurry image—one that is of little value. An ideal microscope is one that has reasonable magnification and good resolution—but, if a choice has to be made between the two, it is always in favor of the resolution.

The resolving power of a microscope depends upon both the wavelength of light that passes through the system and the numeric aperture (NA), which describes the light-gathering ability of the lens system. Some light is lost as it passes from the illumination system to the lens system due to refraction which occurs as the light rays travel from glass to air. The greater the amount of light lost due to refraction, the lower the numerical aperture of the microscope and therefore the lower its resolving power. The refraction of light can be minimized, and therefore the resolving power of the microscope can be maximized, by the use of immersion oil. Immersion oil has approximately the same refractive index as glass. This prevents the refraction of light as it leaves the glass slide, and gives us a clearer image—better resolution—of the magnified object.

FOCUSING

Modern compound microscopes focus on the subject by moving either the lenses or the stage, in relation to each other. On some microscopes the stage moves up and down; on others the body of the microscope moves. In either case, there is always a *course adjustment knob* and a *fine adjustment knob.* Adjustment is complete when the subject is in sharp focus. A competent microbiologist will alternately adjust the

focus and the condenser to achieve maximum resolution.

The various parts of the microscope, as illustrated in **Figure 1.1,** are summarized in Table 1.1. Be sure to locate each of these parts on your microscope.

ROUTINE CARE OF THE MICROSCOPE

A compound light microscope of reasonably high quality costs more than a thousand dollars. They are precision instruments and do require careful handling. The following rules should be considered mandatory:

Carrying the microscope. The microscope is *never carried with one hand.* When you transport the microscope from its cabinet to your work area, one hand supports the microscope from below the base, while the other hand grasps the body.

Placement of the microscope. The ocular lenses are usually held in place by gravity. Do not tilt the microscope in any way beyond that amount of tilt designed into inclined microscopes. If the scope is too high, raise your stool. Never place the microscope on the edge of the lab bench. Place it as far back from the edge as possible.

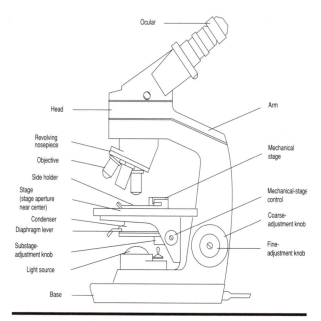

Figure 1.1   The parts of the microscope.

Table 1.1	The parts of the microscope and their functions.
Ocular	Magnifies the image, usually 10x or 15x.
Head	Holds the ocular(s)
Arm	Holds the head and stage
Revolving nosepiece	Rotates the objective lenses into position; usually contains low power, high dry, and oil immersion lenses
Objectives	Magnifies the image—low power (10x), high dry (43x), and oil immersion (97x)
Stage	Holds the slide
Slide holder	Secures the slide on the stage
Mechanical stage	Includes the slide holder and allows for the movement of the slide on the stage
Mechanical-stage control	Moves slide on the stage
Coarse-adjustment knob	Rapidly brings images into focus
Fine-adjustment knob	Slowly brings image into focus
Iris diaphragm level	Controls amount of light entering stage aperture
Substage condenser	Focuses light on specimen
Condenser-adjustment knob	Raises and lowers condenser
Light source	Illuminates specimen
Base	Supports microscope

Peeking and poking. Perhaps the greatest enemy of a microscope (otherwise in good condition) is dust and dirt. Every time you remove a lens or any other part of the microscope, dust can enter the light path. You must resist any temptation to look inside the microscope because when you do, the inner surfaces of lenses and prisms can become dirty and dusty. Microscopes are very difficult to clean. If you have a problem, ask your instructor for help—*do not try to fix the problem yourself.*

Cleaning of lenses. *Always use lens paper for cleaning lenses.* Do not use any kind of tissue, paper toweling, or cloth. The lenses should be cleaned while they are attached to the microscope, by wiping lightly with a circular motion. The circular motion has the effect of moving particles outward to the edge of the lens, where they are removed. If the lens paper

does not clean the lenses, ask the instructor if you should use lens-cleaning solvent. Clean all lenses—objective, ocular, condenser, and any in the light source. Avoid using the same part of the lens tissue for more than one lens, and clean the oil immersion objective last. *Always clean the lenses before and after you use the microscope.*

Storage. The microscope should always be stored in the microscope cabinet. This provides security and allows storage in a relatively dust-free area. Always use the dust cover when one is provided.

"Your microscope." You should get into the habit of using the same microscope each time. You will become familiar with the instrument and any of its idiosyncrasies. Any skilled technician is much more comfortable with his or her own tools.

USE OF THE MICROSCOPE

1. Position the microscope in front of you where you can comfortably look into the ocular lens(es).

2. Turn on the light source and position the low-power objective in the light path.

3. Move the substage condenser all the way up, then back it down about one-quarter inch. Be sure that the iris diaphragm is completely open.

4. Look into the ocular lens(es) and position yourself at the point where you see a full field of view. You may need to move your head back and forth, relative to the lenses, to locate this position. It is important that, if using a monocular microscope (one with only one ocular), you keep both eyes open. If you have a binocular microscope:

    a. Adjust the ocular lenses to match the distance between your eyes. As you look into the lenses, pull the lenses apart or push them together until you see a single round field of view. Move them back and forth so you will recognize when they are not in adjustment.

    b. The lenses of the binocular head must also be adjusted to compensate for the differences in focusing for each of your eyes. Notice that one lens is adjustable and one is not. When you have a slide on the microscope, you should focus it as sharply as possible as you look through the nonadjustable lens.

c. After the subject is in sharp focus with the nonadjustable lens, turn the adjustable lens until the image is sharp in both lenses. Now, when you use both objective lenses, the image should be in sharp focus for both eyes.

5. Place a prepared slide on the microscope stage and center the slide over the hole in the stage. Be sure that the slide is securely held in place by the slide clips or mechanical stage.

6. Using the coarse-adjustment knob, position the low-power objective about one-half inch above the slide.

7. As you look through the microscope, focus on the specimen. Use the coarse adjustment knob to get the specimen into view, then bring the image into sharp focus by using the fine-adjustment knob.

   A word of caution: Most, but not all, modern microscopes are designed so that it is not possible to touch the slide with the lens (referred to as an *autostop*). You should develop the habit of positioning the lens so that any focusing adjustments *will move the lens away from the slide.*

8. If the light is too bright, reduce the intensity of the light source. This may be done in one of two ways: either reduce the voltage to the light by turning a small knob on the light source housing, or close the iris diaphragm located on the light source housing by sliding the iris diaphragm lever. It will be necessary to re-adjust the light intensity when you change lenses.

9. The substage condenser is adjusted by moving it up and down until the sharpest-appearing image is attained. Only very small adjustments in the position of the condenser should be needed. You may also need to adjust the iris diaphragm on the condenser until the image appears free of glare. It is important to learn the difference between real improvements in resolution and apparent improvements caused by increasing contrast. Experiment with the condenser adjustments and be sure to ask for help from your instructor. It will also be necessary to readjust your condenser when you change lenses.

10. After you are satisfied that you have the low-power objective in clear focus, rotate the high dry objective over the object.

    Another word of caution: Modern microscopes are *parfocal*. That is, they are designed so that when one lens is in focus (e.g., the low-power objective), the other lenses (e.g., the high dry objective) will be very near to focus when positioned over the specimen and should require only minor adjustments with the *fine-adjustment knob.*

    The lenses are also designed so they can be rotated into position without hitting the slide. You don't need to move the lenses away from the stage to rotate another lens into position.

    Although most microscopes do have these features, never assume this to be the case. Always watch from the side as you rotate the nosepiece.

11. You will need to focus the lens, but only fine-adjustment should be necessary. (Use the fine-adjustment knob only.) Some adjustments will also be needed to increase the amount of light. Adjust both the iris diaphragm and the condenser. As before, you should make adjustments to produce the sharpest possible image.

12. After you have carefully focused the high dry objective, you may proceed to the oil immersion objective. Remember that you do not need to change focus or move the lenses away from the slide if the microscope is parfocal.

13. Rotate the nosepiece so that the high dry lens is moved out of position, about halfway to the next position. Place one or two drops of immersion oil on the center of the slide. Continue rotating the nosepiece so that the oil immersion objective is centered over the slide. The tip of the lens should be in the oil.

A final word of caution: The space between the lens and the glass slide decreases with increased magnification. A typical oil-immersion lens with a magnification of 97x will be in focus when it is between 0.1 and 0.2 mm above the specimen. The working distance is very small. *When using the oil immersion objective, only use the fine-adjustment knob.* If you cannot focus with the fine-adjustment knob, ask for help.

14. As with the high dry lens, adjustments will be needed to increase the intensity of light and to increase resolution.

WHEN YOU ARE FINISHED WITH THE MICROSCOPE

1. Using the coarse-adjustment knob, move the stage away from the lens. Remove the slide from the stage. If oil was used, carefully wipe the oil from the slide.

2. Use only lens paper to clean the lenses of the microscope.

3. Using a clean piece of lens paper, wipe the ocular, condenser lens, low and high dry lenses with the lens paper. Lastly, remove any oil from the oil immersion lens.

4. Position the nosepiece of the microscope so the low power lens is over the center of the stage. Be sure the microscope is in the upright position. Wrap the electric cord securely around the base of the microscope so that the microscope will not be resting on the cord during storage. Cover the microscope with its dust cover.

5. Return the microscope to it storage cabinet, carrying it with both hands. Be careful not to hit the cabinet or shelf above with the ocular.

HELPFUL HINTS

1. When you change lenses, move the condenser closer to the objective lens.

2. As you increase magnification, the iris diaphragm will need to be adjusted.

3. As you increase magnification, your working distance between the tip of the objective and the slide will decrease markedly.

4. As you increase magnification, adjustment of the microscope will become more critical.

5. As you use the microscope, try to remember the relationship between resolution, substage condenser, and light source.

LABORATORY OBJECTIVES

Your ability to effectively use the microscope is essential to your study of microbiology. In this exercise you will:

- Utilize all powers of magnification with your compound light microscope.

- Explain how the magnification of a microscope is determined and how the immersion oil allows greater resolution at higher magnification.

- Explain the proper care of your microscope.

- Observe microorganisms using the various degrees of magnification available with your light microscope.

MATERIALS NEEDED FOR THIS LABORATORY

1. Microscope

2. Prepared slides:
   Letter 'e'
   Stained human blood smear
   Bacteria, 3 types mixed

3. Immersion oil

4. Lens paper

LABORATORY PROCEDURE

A. USE OF THE LOW-POWER OBJECTIVE

1. Obtain your microscope, set it up, and clean all lenses.

2. Place a letter 'e' slide securely on the stage of your microscope.

3. Follow steps 1-9 in Use of the Microscope (pages 3-4). After the slide is in

clear focus, and while looking through the ocular, adjust the various light controls (the voltage control, iris diaphragm, and condenser—as available on your microscope). What effect do they have on the viewed image?

4. Move the slide on the stage using the mechanical-stage control if available. If your microscope is not equipped with a mechanical stage, carefully move your slide by hand. What happens when you move the slide from left to right? From front to back? Record your observations on the Laboratory Report Form.

5. Remove the slide from the stage and return it to its appropriate place.

B. USE OF THE HIGH DRY OBJECTIVE

1. Obtain a stained human blood smear. Following the procedures above, focus it on low power.

2. After observing it on low power, rotate the nosepiece of your microscope until the high dry lens is in position over the slide. Follow steps 1-11 in Use of the Microscope (pages 3-5) to help you properly focus the slide.

3. Use the fine-adjustment knob to focus on the blood cells. Be sure to adjust your light for maximum resolution. Also, note the amount of adjustment you did with the fine-adjustment knob and which way you needed to turn it. You will find that you will always need to adjust your microscope in the same manner. This is one reason why it is beneficial to use the same microscope consistently.

4. Move the slide so that you see the various forms of human blood cells. Use **Figure 1.2** to help you identify the cells you see. Draw some representative cells on the Laboratory Report Form.

5. Remove the slide from the stage and return it to its appropriate place.

C. USE OF THE OIL-IMMERSION OBJECTIVE

1. Obtain a stained bacteria slide, containing all three shapes of bacteria, mixed.

2. Following the above procedures, focus on low power. Next, take your microscope up to the high dry lens. Be sure to carefully adjust your light, as well as the fine-adjustment knob, as you change lenses.

3. Very carefully, follow the steps 13-14 in Use of the Microscope (page 5) as you go from the high dry to the oil immersion lens. Be sure to adjust your light after changing lenses.

4. On the slide, locate each of the basic bacterial shapes (spherical, rod, and spiral). Draw a few representative cells of each on the Laboratory Report Form.

5. When you have completed looking at your slide, move the stage away from the lens and remove the slide. Carefully blot off any oil from your slide and return it to the appropriate place.

6. Carefully clean the lenses of your microscope with lens paper, being especially careful to remove all of the oil from the oil immersion lens. Follow the steps in When You Are Finished With the Microscope (page 5).

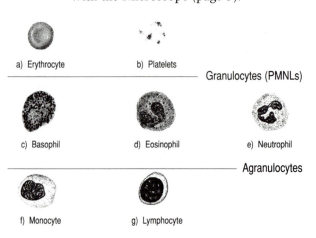

Figure 1.2   Human blood cells.

LABORATORY REPORT FORM

EXERCISE 1
INTRODUCTION TO MICROSCOPY

What was the purpose of this exercise?

A. USE OF THE LOW-POWER OBJECTIVE

1. When you moved the microscope slide from right to left, which way did the image appear to move?

2. Draw the letter 'e' slide as viewed without the use of the microscope.

3. Draw the letter 'e' slide as viewed with the low power lens.

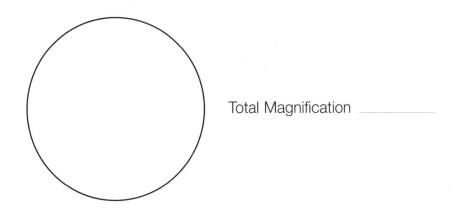

Total Magnification _____

B. USE OF THE HIGH DRY OBJECTIVE

1. Is your microscope parfocal or nearly so?

2. Draw several representative human blood cells as observed through the high dry lens.

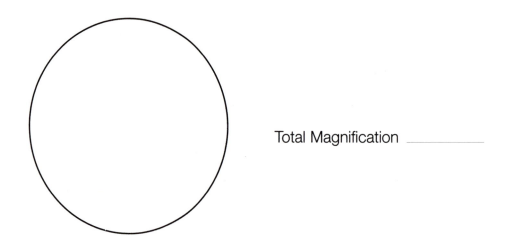

Total Magnification _____

C. USE OF THE OIL IMMERSION OBJECTIVE

1. What light adjustments did you need to perform when you went from high dry to oil immersion?

2. Draw representatives of each of the three shapes of bacteria.

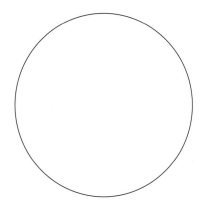

Total Magnification _____

QUESTIONS:

1. Define:
    a. Resolution

    b. Parfocal

2. How is the total magnification of a microscope determined?

3. What is the purpose of the immersion oil?

4. Why can't you view viruses with your microscope?

5. You would like to view the bacterial cells so they appear larger than possible with your micro-
    scope. Why can't you simply increase the magnification?

6. From the background material of this lab, you learned that good compound light microscopes
    often cost around one thousand dollars, yet you have seen microscopes in the toy store that claim
    to have a 10x low power lens, a 43x high dry lens, and a 97x oil immersion lens, and a 10x
    ocular—and they are much cheaper. However, if you were to use one of these microscopes you
    would not be able to see the bacterial slide very well at all. Why not?

PREPARATION OF SMEARS, SIMPLE STAINS, AND WET MOUNTS

BACKGROUND

If we want to learn about individual microorganisms, we must be able to examine single cells. To do this, we have to stain them and view them through a microscope.

BACTERIAL CELL MORPHOLOGY: FOUR BASIC CELL SHAPES

- *Cocci* are spherical cells. Some cocci, especially those that typically form pairs, may be somewhat flattened on the adjacent sides.

- *Bacilli,* or rods, are cylindrically shaped, straight cells that appear rigid. Typically, but not always, the diameter of the cell is constant throughout its length.

- *Vibrios* are shaped like curved rods; their curvature is always less than a half-circle. As with the bacilli, the diameter is generally constant over the length of the cell.

- *Spirilla* are also curved cells; however, their curvature exceeds that of a half-circle. They are characterized as spirals rather than vibrios, and may be classified as either spirilla or spirochetes, depending on their exact structure.

VARIATIONS IN CELL MORPHOLOGY AND CHARACTERISTIC CELL GROUPINGS

Two additional morphological characteristics have proven to be helpful in clinical applications. These characteristics include variations in both cell shape and in the clustering of cells after they divide. Although most bacteria separate from each other after division is complete, some remain together. Often, this tendency to remain together produces cell groups characteristic of the species or genus.

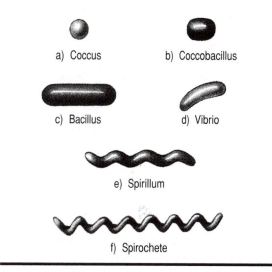

a) Coccus   b) Coccobacillus

c) Bacillus   d) Vibrio

e) Spirillum

f) Spirochete

Figure 2.1   The most common bacterial shapes.

The Cocci: Characteristic Cell Groupings

Cocci, like all bacteria, divide by transverse binary fission. Unlike other bacteria, however, some cocci can vary the plane of division and produce clusters of cells that may take the form of long strands, pairs, packets of four or eight, or random clusters. Since these groupings are characteristic for the genus, they do have diagnostic value and are immediately recognized by the experienced microbiologist.

When cells divide along only one plane and remain in pairs, they are referred to as *diplococci.* If they form filaments several cells long, they are considered *streptococci.*

Neisseria and *Branhamella* are genera of cocci that characteristically form diplococci. These two genera typically have flattened adjacent sides, so much so that they resemble kidney beans arranged with the flattened sides adjacent to each other. The genera *Streptococcus,* *Enterococcus,* and *Lactococcus* are readily identified by the distinctive chains of cocci present when grown in broth cultures.

The *Staphylococcus* and *Micrococcus* typically form random clusters of cells that have been likened to grape-like clusters. This is due to random planes of division during binary fission. The distinction between the staphylococci and streptococci, both frequently encountered human pathogens, is frequently used as an important part of the diagnostic protocol for the cocci.

Some of the cocci divide in alternating planes and produce packets of four cells—*tetrads,* or eight cells—*sarcinae.* These groupings are formed when the plane of each division is at right angles to the previous one.

The Bacilli: Characteristic Cell Groupings

The bacilli only divide in one plane. This can result in the formation of pairs of cells arranged end-to-end—*diplobacilli,* chains of cells arranged end-to-end—*streptobacilli,* or cells arranged side-by-side—*palisade.*

The diphtheroids, which include the *Corynebacterium,* are a prominent component of the normal flora of the skin and upper respiratory tract. Some members of the genus are pathogenic. Diphtheroids have a distinctive morphology due to their tendency to remain attached to each other after division. They do not, however, simply form streptobacilli, but rather snap together, producing clusters of cells that have been likened to "palisades" or

"Chinese characters." In addition, the cells are often thicker at one end and appear to be club-shaped. The combination of palisade or Chinese-character clusters and club-shaped cells is very distinctive. An experienced microbiologist is able to recognize diphtheroids almost immediately.

The Vibrios and Spirillum

The vibrios and spirillum are usually seen as single cells. The major distinctions between them are based on the degree of curvature they exhibit.

The examples given here are not exhaustive. They do, however, represent examples that have significance in clinical situations. See **Figure 2.2**

STAINING PROCEDURES

A chemical to be used as a stain for biological material must have at least two properties: it should be intensely chromogenic (colored) and it must react with some cellular component.

When biological stains which have these properties are used correctly, they enable us to visually examine cells with a considerable level of resolution and definition.

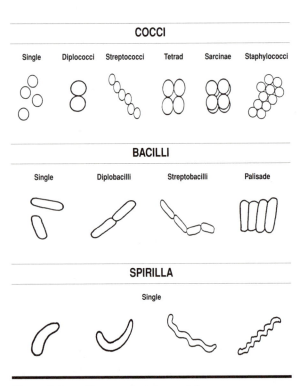

Figure 2.2   Characteristic bacterial arrangements.

Simple stains are ones that can be used as general, all-purpose stains. They are usually basic dyes (a salt with the color in the positive ion) which will stain the cell membranes of most bacteria (a negatively charged structure). It is usually necessary to expose the cells to the stain for only a short period of time during which the positively charged stain will bind with the negatively charged cell membrane. The excess stain is then removed by rinsing before examining the cells with a microscope. In Exercise 3 we will see how selective reactions can be used to differentially stain certain types of cells and/or parts of cells.

PREPARATION AND STAINING OF SMEARS

The preparation of a microscope slide with stained biological matter on it is a two-step process.

1. **Preparation of smears:** A smear is a slide that has had bacteria placed on it and which has been treated to cause the bacterial cells to adhere to the slide.

2. **Staining of smears:** Once the bacteria have been adhered to the slide, they must be stained. This involves exposing the cells to the stain, allowing them to react with it for the required period of time, washing the excess stain from the slide, and then drying it.

UNSTAINED PREPARATIONS: WET MOUNTS AND HANGING DROPS

It is often necessary to examine cells while they are still alive. For example, one method for the determination of bacterial motility requires that we watch the bacteria "swim." By directly observing bacterial movement, it is possible to distinguish between those organisms which are actively motile and show purposeful movement and those which only exhibit Brownian movement, or movement due to molecular bombardment. Brownian movement can determined by the presence of irregular, jerky movement. There are two methods by which we can microscopically observe this.

1. **Wet mounts:** A wet mount is prepared by simply placing a loopful of broth on a slide and then inverting a coverslip onto it so that the drop of broth is between the slide and the coverslip. If the preparation is to be examined over a long period of time, petroleum jelly may be used to seal the edges and prevent drying.

2. **Hanging drop:** If a drop of culture is placed directly on a coverslip, it may be carefully placed on a slide that has a depression molded into it, allowing the drop of broth to hang in the space of the depression, thus the name *hanging drop*. This technique is useful when relatively thick cells, such as protozoans, must be examined.

ASEPTIC TECHNIQUES

Bacteria can be isolated from almost any source; because they exist virtually everywhere, we must be careful not to contaminate our cultures with bacteria from the air or instruments. Obviously, neither do we want to contaminate our work area or anything in it (including ourselves!) with the bacteria in the culture we are studying.

The precautions used to reduce the chances of contamination are collectively referred to as *aseptic techniques.* If these precautions are properly used, the chance of contamination can be reduced to almost zero.

Your laboratory instructor will demonstrate these precautions as they are encountered. They are important and essential for your safety as well as for that of others in your laboratory. Of course, they are also important to prevent you from identifying a contaminant rather than an unknown or to keep you from wasting your time identifying a contaminant instead of the significant organism from a clinical sample. These precautions include:

- Flaming of inoculating loops and needles before and after use.

- Prevention of splattering caused by using needles and loops that have not properly cooled after flaming. The most common source of laboratory contamination and infections are the aerosols created by splattering.

- Flaming the mouths of test tubes and other culture vessels to prevent the inflow of organisms in the air.

- Maintaining the sterility of glassware.

- Proper disposal of contaminated instruments, pipettes, and cultures.

- Proper disinfection of work area before and after use.

- Following designated safety precautions in cases of accidental spills.

- Proper washing of hands with a germicidal soap both at the beginning and completion of the laboratory exercise as well as any time spills occur.

LABORATORY OBJECTIVES

The visual examination of bacterial cells is an important part of their characterization. It is important that you know how to do this. In this exercise, you will:

- Prepare bacterial smears and wet mounts for microscopic study.

- Stain bacterial smears and observe cell morphology and typical groupings.

- Understand the basic principles underlying staining procedures.

- Distinguish between motility and Brownian movement.

MATERIALS NEEDED FOR THIS LABORATORY

A. SMEAR PREPARATION

1. Twenty-four-hour cultures of:
   Staphylococcus epidermidis (slant and broth)
   Escherichia coli (slant and broth)
   Bacillus subtilis (slant)
   Corynebacterium xerosis (slant)

2. Glass slides: The slides must be carefully cleaned and free from residual oils.

B. SIMPLE STAINING

1. Staining reagents:
   Crystal violet
   Methylene blue
   Safranin

2. Prepared slides: It is often helpful for students to use prepared slides as they

familiarize themselves with the appearance of stained bacterial cells.

C. WET MOUNT

1. Coverslips
2. Glass slides
3. Petroleum jelly

LABORATORY PROCEDURE

Aseptic techniques must be used throughout this exercise. This is especially important with the wet mounts because the bacteria are still alive. *Follow all instructions carefully and pay special attention to your instructor when these techniques are demonstrated for you.*

The stains used in this exercise will also stain you fingers and clothing. *Use appropriate caution.*

A. SMEAR PREPARATION

1. Prepare your slides. You will need 3 slides for each stain to be used. Two slides will be needed for the wet mounts.

   a. The glass slides must be especially clean. If there is any residual grease on the slide, the smears will be irregularly spread over the glass and it will be difficult to observe bacterial morphology. Wash all slides with soap and water. This should be followed by both a water rinse and a final rinse with 95% ethyl alcohol. The slides must be dry prior to their use.

   b. If you are using slides with a frosted end, you can use a pencil to label the slide; the frosted end will serve as a marker to keep the slide properly oriented (right side up and right to left). If frosted slides are not available, use a marking pen or a wax marking pencil for labeling your slide.

   c. Using a marking pen, draw two circles about one-half inch in diameter on each slide as indicated in **Figure 2.3.**

2. Preparation of smears from agar cultures. You will need 1 set of smears for each stain to be used.

   a. Place a small drop of water in each circle on the slide. Your inoculating

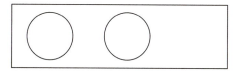

Figure 2.3   Smear preparation.

loop is a handy means of applying the proper amount of water.

b. Using your inoculating loop, transfer a small amount of culture to the drop. Emulsify the culture in the drop, mixing it thoroughly with the drop of water. This will produce a moderately turbid (cloudy), but not opaque, smear. Spread it evenly throughout the circle. Be sure to flame your loop before *and* after use.

c. Repeat until each of the slant cultures have been prepared on a slide.

d. Allow the slide to air dry.

e. Heat fix the bacteria to the slide by passing the slide through the flame of your Bunsen burner several times. If you do not heat the slide enough, the bacterial cells will be washed off during the staining process. If it is overheated, the cells may become carbonized.

3. Preparation of smears from broth cultures. You will need one set of smears for each stain to be used.

a. Use your inoculating loop to transfer a loopful of a broth culture to your slide. Be sure to mix the broth before removing the loopful. Use the aseptic technique and remember to flame your loop before and after use.

b. Spread the drop over the marked area.

c. If the broth culture is not very turbid you may need to apply two or three loopfuls to the slide.

d. Repeat until each of the broth cultures has been prepared on the slide.

e. Allow the slide to air dry.

f. Heat fix as described above.

B. STAINING

1. After the slides have completely cooled, place them on a staining rack located over either the sink or a staining dish or tray.

2. Flood the slides with stain. Make sure each smear is covered completely. Do not allow the smears to dry; add additional stain as needed. Allow the smears to stain according to times shown:
Crystal violet: 30 seconds
Methylene blue: 1 minute
Safranin: 1 minute

3. Rinse the excess stain from the slide by holding it under a stream of gently flowing cold water. Do not allow the stream of water to fall directly on the smear, but let it flow over the smear.

4. Remove excess water from the slide by tapping a long edge of the slide on some paper toweling.

5. Allow the slide to air dry. Some workers prefer to blot the slide with paper toweling or bibulous paper. If your instructor allows this, place the slide on the toweling or paper, stained side up, and carefully blot it. Do not rub the slide with the paper.

NOTE: Staining procedures, especially simple stains, do not always kill the bacteria, especially the spore-formers. Blotting the slide may transfer viable bacteria to the paper. Air drying is the procedure of choice.

6. Examine all the smears with the oil-immersion objective. Record your observations on the Laboratory Report Form.

C. WET MOUNT

1. Prepare a wet mount of each broth culture. Carefully transfer a loopful of culture to the center of a clean slide. Invert the coverslip over the center of the slide.

2. Lower the coverslip onto the slide so that one edge touches first (See **Figure 2.4**). The cover slip can then be lowered onto the slide. This procedure will minimize the formation and retention of bubbles in the broth droplet.

3. If you intend to examine the wet mount for more than several minutes, you should seal the edges with petroleum jelly. Using a toothpick, carefully place a small bead of jelly (about 1.0 mm) around the four sides of the coverslip *before* placing the loopful of broth on the slide.

4. Invert the coverslip and position it as described above. The jelly should form a seal when you lower the coverslip onto the slide. A small amount of pressure may be needed to completely seal the edges.

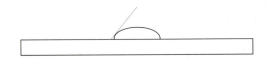

Figure 2.4   Preparation of a wet mount.

5. Examine the wet mount with the high-power and oil-immersion objectives of the microscope. Determine if the bacteria are motile by observing directional movement (as opposed to simple Brownian movement). *Remember that the bacteria are still viable. Dispose of the wet mount in a safe manner.*

6. Complete the Laboratory Report Form for this exercise.

LABORATORY REPORT FORM

Name _____

Section _____

EXERCISE 2
PREPARATION OF SMEARS, SIMPLE STAINS, AND WET MOUNTS

What was the purpose of this exercise?

B. SIMPLE STAIN

	Staphylococcus epidermidis	Escherichia coli	Bacillus subtilis	Corynebacterium xerosis
Appearance				
Morphology				
Arrangement				
Stain Used				
Magnification				

Compare the appearance of the cells when stained with the different stains.

C. WET MOUNT

	Staphylococcus epidermidis	*Escherichia coli*
Appearance		
Motility observed		

QUESTIONS:

1. Why is it necessary to heat fix bacterial smears?

2. What advantages are there to observing unstained bacterial specimen?

3. Distinguish between true motility and Brownian movement.

DIFFERENTIAL STAINING PROCEDURES

GRAM STAIN, ACID-FAST STAIN, AND METACHROMATIC-GRANULE STAIN

Simple stains, as shown in Exercise 2, stain biological materials indiscriminately. Fortunately, by modifying the staining procedure, using special stains, or adding chemicals, it is possible to differentially stain bacteria. In some cases, only certain *types* of cells will be stained; in other cases, only certain *parts* of cells will be stained. These procedures, which allow us to stain different cells or components different colors, are referred to as *differential staining* procedures.

BACKGROUND

Differential staining, unlike simple staining, requires at least three component factors or steps to make the procedure differential. These include:

- The *primary stain* is used to color the "target" cells or cell parts. It is the staining reagent that actually stains the cell or cell component that you want to examine.

- A *mordant* is a chemical that reacts with the primary stain and with the cell or component you want to see. Its purpose is to enhance the retention of the primary stain.

  Selective treatment, on the other hand, is an added step that takes advantage of some special cell characteristic and results in retention of the primary stain by the target cells or cell components. Examples of selective treatment include heating cells or decolorizing with an alcohol solution. *In some procedures both a mordant and a special selective treatment are used.*

- The *counterstain* is usually a simple stain. It is used to stain everything that was not stained by the primary stain. It is generally a contrasting color.

You will use three differential stains in this exercise—the Gram stain, the acid-fast stain, and the metachromatic-granule stain. These staining techniques are among the most often used staining procedures in the clinical microbiology laboratory. They will serve as examples of differential staining procedures, but do not represent an all-inclusive list of the available staining techniques. Other staining procedures that are less commonly used will be discussed in Exercise 4.

THE GRAM STAIN

The Gram stain allows rapid distinction between two major groups of bacteria on the basis of the structure and composition of their cell surface membranes and cell walls. Because of changes that may occur in the bacterial cell wall with age, the Gram stain technique is most reliable when performed on twenty-four to forty-eight hour cultures or on samples directly taken from a patient. When performed on older cultures, Gram-positive organisms often appear to be Gram-negative. The Gram stain procedure is probably the single most commonly used staining procedure in microbiology. When microbiologists are discussing a bacterial species, the gram reaction is almost always given. For example, they may refer to a "gram-positive coccus," or a "gram-negative bacillus." Virtually all diagnostic protocols used in microbiology begin with the determination of the organism's shape and gram reaction.

Primary Stain:	Gram's Crystal violet
Mordant:	Gram's iodine
Selective treatment:	Ethanol/acetone rinse
Counterstain:	Gram's Safranin

The Gram's crystal violet is applied to a smear and allowed to stain the cells for 30 seconds. The crystal violet is then rinsed off the slide with the iodine solution and the smear flooded with additional mordant. After one minute, the smear is rinsed with either ethanol or a mixture of ethanol and acetone. The solvent will remove the crystal violet iodine complex from gram-negative cells only. The counterstain, Gram's safranin, is then used to stain all cells that have not retained the crystal violet.

Gram-positive cells:	Purple
Gram-negative cells:	Pink or red

THE ACID-FAST STAIN

The genera *Mycobacterium* and *Nocardia* are two of a very few genera that stain positively with the acid-fast stain. These bacteria have a waxy component in their cell wall that makes the wall almost impermeable to solutes dissolved in water. The wall can be made permeable by heating. The genus *Mycobacterium* includes two important pathogens: *M. tuberculosis* and *M. leprae*. The presence of acid-fast bacteria in sputum or tissue is considered presumptive evidence of either tuberculosis or leprosy. Thus, the acid-fast stain has significance in medical microbiology.

In the Ziehl-Neelsen procedure, a smear is flooded with the carbol fuchsin stain and heated with a Bunsen burner flame or a steam bath for five minutes. After the slide cools, it is rinsed with acid-alcohol until the dye no longer flows from the smear. The slide is then gently rinsed with water, and counterstained with methylene blue. The acid-alcohol will remove all carbol fuchsin that has not been "trapped" inside acid-fast cells.

The Kinyoun's procedure is often referred to as *cold staining*. The concentration of carbol fuchsin and phenol in the stain has been increased, eliminating the need for heat. The smear is flooded with the

Ziehl-Neelsen's Procedure

Primary stain:	Ziehl-Neelsen's Carbol fuchsin
Selective treatment:	Heating cells and acid-alcohol rinse
Counterstain:	Methylene blue

Kinyoun's Procedure

Primary stain:	Kinyoun's Carbol fuchsin
Selective treatment:	Acid alcohol rinse
Counterstain:	Methulene blue

Kinyoun Carbol Fuchsin and allowed to set for five minutes. It is then rinsed and counterstained exactly as was done in the Ziehl-Neelsen procedure.

In addition to these stains, there is a fluorescent staining technique that is being used increasingly in diagnostic laboratories because of its high diagnostic specificity. This procedure requires the use of fluorescence microscopy and will not be done in our laboratory.

Acid-fast bacteria:	Pink
Non-acid-fast bacteria:	Blue

THE METACHROMATIC-GRANULE STAIN

Metachromatic granules, or volutin, are storage crystals produced by some bacteria when there is excess phosphate and carbohydrate in their growth medium (a special medium is usually needed). The storage crystals are deposits of cellular phosphate and are large enough to see when stained. Although several kinds of bacteria produce these granules, their production is not common, and those bacteria that do so are relatively distinctive. One of the most commonly encountered species that produces the granules is *Corynebacterium diphtheriae*. The detections of diphtheroids with metachromatic granules is presumptive evidence for *C. diphtheriae*.

The metachromatic-granule staining procedure uses a stain that adheres strongly to the phosphate crystals composing the granules. In the Albert's procedure, the smear is covered with the stain and allowed to react for a few minutes. It is then rinsed and counterstained with Gram's iodine. Cells with metachromatic granules will have dark blue to black

Albert's Procedure

Primary stain:	Albert's Stain
Counterstain:	Gram's Iodine

Loeffler's Procedure

Primary stain:	Loeffler's methylene blue
Counterstain:	None

granules in their cytoplasm, which appears uniformly light green.

The Loeffler's procedure requires simple staining with the methylene blue for several minutes. The granules appear dark blue against a light blue cytoplasm.

Positive cells:	**Have visible granules in their cytoplasm**

LABORATORY OBJECTIVES

Differential and selective stains allow the microbiologist to rapidly separate bacteria into smaller, more manageable groups for identification. For this exercise you should:

- Understand the significance of the Gram-stain procedure and know what components of bacterial cells are responsible for the differential staining reactions.

- Understand the clinical importance of the acid-fast and metachromatic-granule staining procedures; know how these staining procedures work.

- Be able to discuss the components of differential staining procedures and why they differ from simple staining procedures.

- Begin to categorize bacteria on the basis of their cellular morphology and staining characteristics.

MATERIALS NEEDED FOR THIS LABORATORY

These procedures may be done during one or more laboratory periods, as designated by your instructor. In addition, your instructor may direct you to perform stains on select organism(s), comparing your results with those achieved by other members of your laboratory section.

A.   GRAM STAIN

1. Twenty-four hour cultures of:
   Staphylococcus epidermidis
   Escherichia coli
   Mycobacterium smegmatis (72-hour culture)
   Corynebacterium xerosis
   Moraxella catarrhalis
   Unknown

2. Gram-stain reagents:
   Gram's crystal violet
   Gram's iodine
   Ethanol or acetone/ ethanol
   Safranin

3. Prepared Gram stains of:
   Staphylococcus epidermidis
   Escherichia coli.

B.   ACID-FAST STAIN

1. Twenty-four hour cultures of:
   Staphylococcus epidermidis
   Escherichia coli
   Mycobacterium smegmatis (72-hour culture)
   Unknown

2. Acid-fast reagents:
   Ziehl-Neelsen's Carbol Fuchsin or
   Kinyoun's Carbol Fuchsin
   Acid alcohol
   Methylene blue

3. Small pieces of paper toweling, approximately 3/4 inches by 2 inches

4. Prepared slide of *Mycobacterium tuberculosis*

C.   METACHROMATIC-GRANULE STAIN

1. Twenty-four hour cultures of:
   Staphylococcus epidermidis
   Escherichia coli
   Corynebacterium xerosis
   Unknown

2. Metachromaticgranule stain reagents:
   Albert's stain and Gram's iodine
   or Leoffler's methylene blue

3. Prepared slide of:
   Corynebacterium diphtheria.

LABORATORY PROCEDURE

The aseptic techniques required in the previous exercise are also necessary here. Use the same precautions to prevent staining of your fingers, clothing, and jewelry.

When differential stains are used, proper controls are needed to ensure that the procedure is working correctly. An easy way to employ such controls is to make sure that there is a positive and negative smear on the same slide (why?).

A. GRAM STAIN

1. Prepare smears of bacterial cultures as indicated by your instructor. Heat fix.

2. Complete the Gram stain, see **Figure 3.1**.

   a. Flood the smears with Gram's crystal violet.

   b. After 30 seconds, remove the crystal violet by rinsing gently with the iodine solution.

   c. Flood the smears with Gram's iodine and allow it to react for one minute.

   d. As you hold the slide over the sink or staining tray, allow the ethanol or acetone/ethanol to flow over the smears. As soon as the color stops flowing from the smear, rinse the slide with water. This step, called *decolorization,* is the most critical. It is easy to use too much alcohol, resulting in poorly stained Gram-positive cells.

   e. Cover the smears with safranin for 1 minute.

   f. Rinse the stain from the slide, drain the excess water, and air dry.

3. Examine all smears with the oil-immersion objective.

4. Examine the prepared stains as directed.

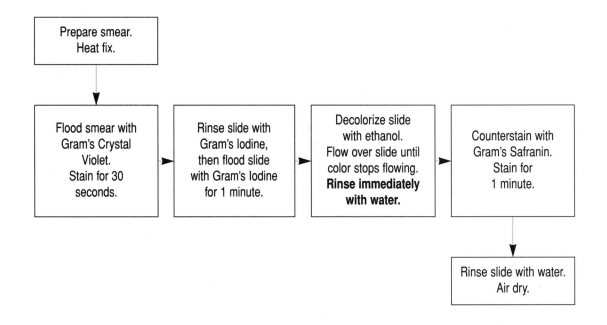

Figure 3.1   Gram stain procedure.

5. Complete the Laboratory Report Form, Part A.

B. ACID-FAST STAIN

1. Prepare smears of bacterial cultures as indicated by your instructor. Heat fix.

2. Complete the acid-fast stain (see **Figure 3.2**) using the procedure indicated by your instructor.

Ziehl-Neelsen's Procedure:

a. Construct a steam bath or use a ring stand to support the slide while it is being heated.

b. Cut or tear a small piece of paper toweling that is larger than the smear, but slightly smaller than the slide (about 3/4 in. by 2 in.).

c. Place the slide on the steam bath or ring stand, cover the smear with the piece of paper toweling, and flood the slide with Ziehl-Neelsen carbol fuchsin.

d. Heat the slide for *not less than five minutes,* If you use a Bunsen burner flame to heat the slide, vapors should appear over the stain. *Do not allow the fluid to boil—do not allow the stain to dry.* As the stain evaporates, add more to the smear.

e. After the smears have been heated for at least five minutes, allow the slides to cool, then rinse with water to remove excess stain. Do not allow the

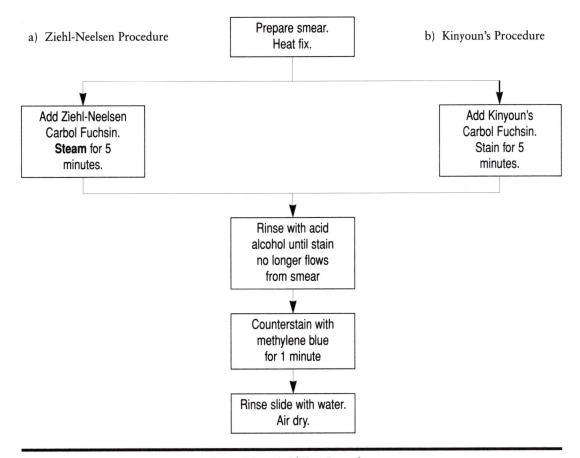

a) Ziehl-Neelsen Procedure

b) Kinyoun's Procedure

Prepare smear.
Heat fix.

Add Ziehl-Neelsen
Carbol Fuchsin.
Steam for 5
minutes.

Add Kinyoun's
Carbol Fuchsin.
Stain for 5
minutes.

Rinse with acid
alcohol until stain
no longer flows
from smear

Counterstain with
methylene blue
for 1 minute

Rinse slide with water.
Air dry.

Figure 3.2   Acid Fast Procedure.

paper to fall into the sink. Use your needle, loop, or forceps to place it in the waste container.

f. Rinse the slide with acid alcohol until stain no longer flows from the smear. The acid-fast stain, while not as rapidly decolorized as the gram stain, can be de-stained if the alcohol is applied for too long a period of time. Rinse gently with water.

g. Counterstain with methylene blue for one minute; rinse, drain and air dry.

Kinyoun's Procedure (Cold Staining):

a. Flood the smear with Kinyoun's carbol fuchsin and allow it to stain for five minutes.

b. Rinse the smear with water to remove excess stain.

c. Rinse the slide with acid alcohol until stain no longer flows from the smear. The acid-fast stain, while not as rapidly decolorized as the Gram stain, can be destained if the alcohol is applied for too long a period of time. Rinse gently with water.

a) Albert's Procedure

b) Loeffler's Procedure

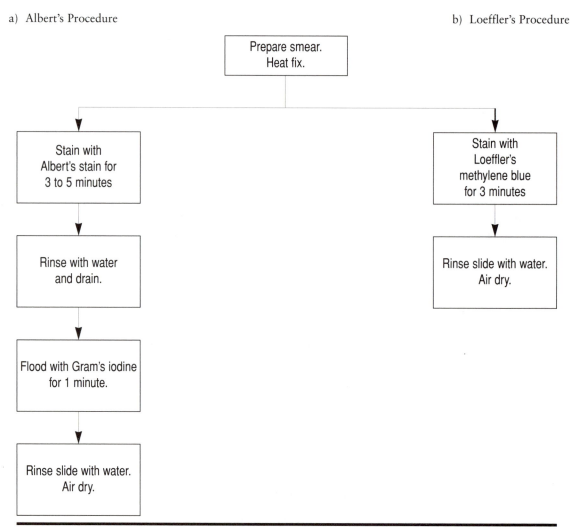

Figure 3.3   Metachromatic Granule Stain.

d. Counterstain with methylene blue for one minute; rinse, drain and air dry.

3. Examine all smears with the oil immersion objective.

4. Examine the prepared slide as directed.

5. Complete the Laboratory Report Form, Part B.

C. METACHROMATIC-GRANULE STAIN

1. Prepare smears of bacterial cultures as indicated. Heat fix.

2. Complete the metachromatic-granule stain (see **Figure 3.3**) using the procedure indicated by your instructor.

   Albert's Procedure:

   a. Flood the smears with Albert's stain for three to five minutes.

   b. Rinse the slide with water and drain.

   c. Flood the smear with Gram's iodine and allow to react for one minute.

   d. Rinse, drain, and air dry.

   Loeffler's Procedure:

   a. Flood the smears with Loeffler's methylene blue for three minutes.

   b. Rinse, drain, and air dry.

3. Examine all smears with the oil immersion objective.

4. Examine the prepared slide as directed.

5. Complete the Laboratory Report Form, Part C.

NOTES

LABORATORY REPORT FORM

EXERCISE 3
DIFFERENTIAL STAINING PROCEDURES

What was the purpose of this exercise?

A. GRAM STAIN

ORGANISM			
APPEARANCE			
MORPHOLOGY			
GRAM REACTION			
MAGNIFICATION			

ORGANISM			
APPEARANCE			
MORPHOLOGY			
GRAM REACTION			
MAGNIFICATION			

B. ACID FAST STAIN

Which procedure did you use?

ORGANISM			
APPEARANCE			
MORPHOLOGY			
GRAM REACTION			
MAGNIFICATION			

C. METACHROMARTIC-GRANULE STAIN

Which procedure did you use?

ORGANISM			
APPEARANCE			
MORPHOLOGY			
GRAM REACTION			
MAGNIFICATION			

QUESTIONS:

1. A mixed smear of *Staphylococcus epidermis* and *Escherichia coli* is Gram stained. Describe how it would appear under the following conditions:

    a. Properly stained smear made from 24-hour cultures.

    b. Properly stained smear made from 7-day old cultures.

2. Which, if any, step could be omitted in the Gram stain procedure? Why?

3. In performing the Acid-Fast stain, a student grabs ethanol/acetone instead of acid alcohol for decolorizing. What results would the student obnserve after completing the staining procedure? Why?

4. Why should controls be stained on the same slide as an unknown smear?

5. What is the significance of a positive stain for metachromatic granules?

ADDITIONAL DIFFERENTIAL STAINING PROCEDURES
DEMONSTRATION OF BACTERIAL CYTOLOGY

There will be four staining techniques demonstrated in this exercise. The exercise is structured so that each of them can be completed independently. Your laboratory instructor will indicate which of the procedures you should complete.

The staining procedures covered in Exercise 3 (Gram stain, acid-fast stain, and metachromatic-granule stain) are commonly used for the identification of clinical isolates. However, they are only three of many such procedures. The procedures that will be covered in this exercise, while not as commonly used in routine laboratory work, are useful because they stain cellular organelles that would not otherwise be visible. Their diagnostic value is somewhat less than the three other procedures, but their value for the study of bacterial cytology is greater.

The structures that are revealed by these procedures are, for the most part, subcellular organelles and are smaller than the cells that contain them. You will need to pay special attention to getting maximum resolution from your microscope. Examination of prepared slides would be quite helpful for reviewing your microscope and staining techniques.

BACKGROUND

A. NEGATIVE STAIN

The *negative stain,* as the name implies, stains the background but not the cell itself. Its effect is like that of a photographic negative—everything appears dark except the structure you want to examine. The negative staining procedure is frequently used to examine the capsules of bacteria, but it can be used to stain virtually any structure that is impermeable to the stain. (We will discuss specific capsule stains in a later part of this exercise.)

The most commonly used procedure is mixing the bacteria with India ink or Dorner's nigrosin and examining the resulting suspension as a wet mount (see Exercise 2) or as a very thin smear. India ink is a colloidal suspension of carbon particles. When bacteria are mixed with India ink, the carbon particles cannot penetrate the surface layers of the cell; the cell remains unstained while the background appears dark. Nigrosin is an acidic stain and is repelled by the cell membrane, also resulting in the cell remaining unstained against a dark background.

B. CAPSULE STAIN

The *capsule* or *slime layer* is a deposit of slippery, mucoid material that forms around many bacterial cells. Some forms produce capsules that are very small, while others, such as *Klebsiella pneumonias* and *Streptococcus pneumoniae,* produce prominent capsules. Many studies have linked the capsule with enhanced bacterial virulence, and the capsule has played a fairly prominent role in the history of microbiology. (Consult your text for information about capsules, bacterial transformation and the Quelling reaction.)

Although it is possible to see the capsule with negatively stained smears, several techniques can be used cells. In one procedure, a negatively stained thin smear is gently heated and then lightly stained. In to another procedure, known as the Anthony method, the capsules are coated with copper sulfate to make them appear light blue.

C. SPORE STAIN

Bacterial endospores are heat-resistant structures that enable the survival of endospore-forming bacteria when exposed to normally lethal temperatures or from many types of disinfectants and harmful environmental conditions. For example, endospores are not affected by prolonged boiling. As a result of this high resistance to heat killing, suspensions of bacterial endospores have become the accepted standard for monitoring the effectiveness of most sterilization procedures. The most commonly encountered endospore-forming bacteria belong to the genera *Bacillus* and *Clostridium*, although at least one coccus is known to make endospores. Because there are very few endospore-forming organisms, the presence of endospores can be very diagnostically significant.

The morphology and location of the spore, relative to the cell that produced it, are quite varied. These variations are used to help classify the bacilli and the clostridia. When a spore-forming bacterial species is described, the following characteristics of the spore are usually noted:

Shape:
  Oval
  Round

Diameter:
  Larger than the cell
  Smaller than the cell
  About the same diameter

Location:
  Central (in the middle)
  Terminal (at the very end)
  Subterminal (in between)
  Lateral (off the long axis)

The wall of the endospore is impermeable to water—as well as to anything dissolved in water. However, heating the cells in the presence of a stain allows sufficient dye to enter the spore so it can be seen. The procedure used is similar to that used in the Ziehl-Neelsen procedure to stain acid-fast bacteria (see Exercise 3).

D. FLAGELLA STAIN

Bacteria that produce *flagella* are motile. A positive test for motility by the wet-mount procedure or by the use of motility medium is evidence that the bacterium being tested has flagella. Unfortunately, however, simply knowing whether or not a bacterium has flagella is not enough.

In many protocols for the identification of the gram-negative rods, serological methods are used to test the antigenic nature of the flagella. In other cases it is important to determine the location and number of flagella on a single cell.

The location and sometimes the number of flagella on bacterial cells have been found to be consistent for any given species and may be used to help classify bacteria into their correct taxons. If they are polarly flagellated, do they have one flagellum at one end, one at both ends, or a tuft of flagella at one or both ends?

Flagella have diameters considerably smaller than the resolving power of the best optical instrument, even when it is used correctly. The flagella stain makes these organelles visible by coating them with sufficient amounts of mordant (tannic acid and alum) to make them visible. Two procedures are common: Gray's method and Leifson's method. Gray's method does not require special dyes, but it is somewhat more touchy in terms of getting excellent results. In any case, both procedures require that: (a) the slide be absolutely clean; (b) the cells be handled in an exceptionally gentle manner to avoid breaking off the flagella; and (c) capsular material be removed before staining.

LABORATORY OBJECTIVES

- Understand the nature of the bacterial endospore and its relationship to survival of the organism under adverse conditions.

- Understand the nature of the capsule and its relationship to virulence.

- Understand the nature of bacterial flagella and their role as organelles of locomotion.

- Gain additional experience with high-resolution microscopy and with additional cytological techniques.

MATERIALS NEEDED FOR THIS LABORATORY

These procedures may be performed during one or more laboratory periods, as designated by your instructor.

A. NEGATIVE STAIN

1. Twenty-four hour cultures of:
    Proteus vulgaris
    Pseudomonas aeruginosa
    Lactococcus lactis

2. India ink (high quality) or
    Dorner's nigrosin

3. Glass slides and coverslips

4. Tubes containing approximately
    2.0 ml saline

5. Prepared slides stained with
    the negative stain

B. CAPSULE STAIN

1. Twenty-four-hour cultures of
    Klebsiella pneumoniae
    Streptococcus pneumoniae

2. India ink (high quality) or
    Dorner's nigrosin

3. Crystal violet (1%, wt/vol, aqueous)

4. Copper sulfate with 5 waters of
    hydration (20%, wt/vol, aqueous)

5. Prepared slides stained with capsule stain

C. ENDOSPORE STAIN

1. Seventy-two hour cultures of:
    Bacillus cereus
    Bacillus megaterium
    Bacillus subtilis
    Clostridium sporogenes

2. Malachite green (0.5%, wt/ vol, aqueous)

3. Gram's safranin

4. Prepared slides stained with spore stains

D. FLAGELLA STAIN

1. Twenty-four-hour cultures of:
    Proteus vulgaris
    Pseudomonas aeruginosa

2. Carefully cleaned slides

3. Saline solution to suspend agar cultures.

4. Gray's solution "A." This solution
    should be mixed just prior to use. It
    does not store well and must be filtered
    just prior to usage.

5. Ziehl's carbol fuchsin

6. Prepared slides stained with flagella stain

LABORATORY PROCEDURE

A. NEGATIVE STAIN

WET MOUNT

1. If organisms are growing on an agar
    slant, transfer a loopful of culture into
    about 2 ml of saline. If they are in
    broth, you do not need to suspend
    them in saline.

2. Place one loopful of India ink or
    Dorner's nigrosin on the slide. Place a
    loopful of the bacteria suspension next
    to, *but not touching,* the drop of India
    ink. *Do not mix the drops at this time.*

3. Carefully lower a coverslip over the
    drops. Allow it to cover both the stain
    and the suspension of bacteria. When
    the coverslip is in place, the drops will
    have mixed in such a way that a
    stained gradient will be formed. See
    Figure 4. 1.

Figure 4.1   Preparation of a negative stain wet mount.

4. If a sealed wet mount is desired, place
    the droplets on the coverslip and use
    the procedure given in Exercise 2.

5. Examine the wet mount with the
    low-power and the high-dry objective
    lenses. Find the part of the slide that
    has the optimum contrast for observ-
    ing both capsules and the refractile
    bacterial cell.

6. The wet-mount staining procedure
    does not subject the cell to drying. Any
    artifacts that occur when cells are dried
    will be avoided.

7. Record your observations on the Lab-
    oratory Report Form, Part A.

THIN SMEAR

1. Use the same bacteria as before.

2. Place one loopful of india ink or Dorner's nigrosin near the end of a glass slide. Mix a loopful of bacterial suspension with the stain. Try to keep the droplet as small in diameter as possible.

3. Hold a short edge of another glass slide against the first slide, at an angle of about 45 degrees. Beginning near the center of the slide, "back" the tilted slide up until it touches the mixture of stain and bacteria. At this point the mixture should spread out along the edge of the slide. See **Figure 4.2.**

4. Quickly push the tilted slide forward until it clears the slide that the smear will be on. If the slide was clean and your pushing was smooth, the droplet will have smeared across the slide, producing a thin smear with a feathered edge. Allow the slide to air dry.

5. Observe the organisms with all three objective lenses, starting with the low power lens. The best part of the slide is usually near the feathered edge. The cells should be distinctly visible as unstained structures in a smooth-textured dark background.

6. Record your observations on the Laboratory Report Form, Part A.

NOTE: This staining procedure does not necessarily kill the bacterial cells. Be sure to soak the slides in a disinfectant following observation.

B. CAPSULE STAIN

THIN SMEAR PROCEDURE

1. Prepare a negatively stained thin smear as described above.

2. Gently heat fix the smear.

3. Flood the smear with crystal violet for about one minute.

4. Gently wash the slide with cold water. Air dry.

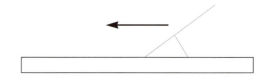

Figure 4.2   Preparation of a negative stain thin smear.

5. Examine the slide with all three objective lenses. The cells should be stained dark blue to purple and should be approximately in the center of the unstained capsule. The background is, as before, darkly stained.

6. Record your observations on the Laboratory Report Form, Part B.

ANTHONY'S METHOD

1. Prepare a smear as described in Exercise 2. Allow the smear to air dry. Do not heat fix.

2. Stain the smear with 1% aqueous crystal violet for about two minutes.

3. Gently wash the slide with a 20% solution of copper sulfate. Drain. Blot dry.

4. Examine the slide with all three objective lenses. The cells should appear dark blue or purple, with light blue capsules.

5. Record your observations on the Laboratory Report Form, Part B.

C. ENDOSPORE-STAINING PROCEDURE (SCHAEFFER-FULTON METHOD)

1. Seventy-two hour cultures of *Bacillus* or *Clostridium* should be used.

2. Prepare a smear of the cultures in the usual manner. Air dry and heat fix.

3. Place a small piece of filter paper over the smear (refer to Ziehl-Neelsen procedure for the acid-fast stain in Exercise 5) and flood it with the malachite green stain.

4. Gently heat the flooded smear by steaming it over a boiling water bath. The stain should steam for at least five minutes. Do not allow the stain to boil. Add additional stain as needed so the smear does not dry.

5. After the slide has cooled, discard the paper in a suitable container and wash the slide with cold water until color no longer flows from the smear.

6. Counterstain with safranin for one to three minutes. Rinse with cold water and air dry.

7. Examine the slides with all three objective lenses. The spores will stain green and the cytoplasm will be pink.

8. Record your observations on the Laboratory Report Form, Part C.

D. FLAGELLA-STAIN PROCEDURE (GRAY'S METHOD)

1. Sixteen- to twenty-four hour broth cultures produce excellent results and may be directly applied to the slide. If it is necessary to use an agar culture, transfer enough loopfuls of culture to produce a distinctly turbid suspension in saline and allow the cells to swim around for about 30 minutes. This will cause the bacteria to lose most of their capsular material.

2. Carefully clean one glass slide for each of the bacteria to be stained. A paste of non-abrasive cleanser can be rubbed on the slide and rinsed in flowing warm water. Either air dry or wipe with a fresh Kimwipe tissue. After the slide has been cleaned, do not touch the part of the slide that will hold the smear.

3. Place one drop of bacterial suspension near the edge of the slide. Tilt the slide and allow the droplet to flow over the surface. Rock the slide to spread out the smear. Do not use you needle or loop to spread the droplet.

4. Allow the smear to air dry.

5. Flood the smear with filtered Gray's Solution "A." Allow the cells to stain for about eight minutes. Some experimentation may be needed to determine the best staining time—between six and ten minutes usually works well.

6. Gently rinse the slide with distilled water. Do not squirt the water directly on the smear, but let the water flow over it until the stain is removed.

7. Cover the smear with a small piece of filter or blotting paper. Flood the smear with Ziehl's carbol fuchsin for three minutes.

8. Remove the paper by lifting (not sliding) it off the smear. Rinse gently with distilled water. Air dry.

9. Examine the slide with all three objective lenses. You may need to search several sections of the slide to find good examples of the flagellated cells. The cells and their flagella will stain red.

10. Record your observations on the Laboratory Report Form, Part D.

NOTES

LABORATORY REPORT FORM

EXERCISE 4
ADDITIONAL DIFFERENTIAL STAINING PROCEDURES

What was the purpose of this exercise?

A. NEGATIVE STAIN

ORGANISM			
APPEARANCE			
MORPHOLOGY			
MAGNIFICATION			

B. CAPSULE STAIN

ORGANISM			
APPEARANCE			
MORPHOLOGY			
MAGNIFICATION			

C. ENDOSPORE STAIN

ORGANISM			
APPEARANCE			
MORPHOLOGY			
SPORE LOCATION			
MAGNIFICATION			

D. FLAGELLA STAIN

ORGANISM			
APPEARANCE			
FLAGELLA ARRANGEMENT			
MAGNIFICATION			

QUESTIONS:

1. Why is the size more accurate in a negative stain than in a simple stain?

2. Can any stain be used for negative staining? Why or why not?

3. How can the presence of a capsule contribute to an organism's virulence?

4. What advantage is it for *Clostridium* to form endospores?

5. List three diseases caused by organisms that form endospores.

    a.

    b.

    c.

6. We can see the presence of spores in simple stained cells. Why do we then have to perform a spore stain?

CALIBRATION OF THE MICROSCOPE

If your microscope is to be used to obtain quantitative data, it is necessary that it be calibrated so that accurate linear measurements can be made. In this exercise we will determine the diameter of the field of view and calibrate an ocular micrometer. In later exercises we will use the calibrated microscopes to determine the sizes of red blood cells, yeast and bacteria, and for cell counting.

BACKGROUND

Most students initially consider the microscope to be a device that helps them gather nonquantitative information about the specimen they are examining. We tend to think in descriptive terms about what we see through the lenses, although we often do make semi-quantitative observations. For example, we might use the term "*small* coccus" or "*long* bacillus" when describing types of cells. We might describe a bacterial spore as being larger in diameter than the vegetative cell that produces it.

The microscope can be used to obtain fairly precise quantitative information about our specimen. This information can take the form of cell measurements, such as diameter and length, or it can be related to the number of cells present in a given sample. When the microscope is properly calibrated, it is possible to use it to measure the size of the cells and to count the number of cells, in the sample.

THE OCULAR MICROMETER

If the microscope is to be used quantitatively, some means of determining relative lengths must be incorporated into the optical path of the lens system. That is, we must do the equivalent of putting a ruler under the microscope. The most frequently used "ruler," or *ocular micrometer,* is a small piece of glass with a graduated scale etched onto its surface. The graduated scale usually goes from zero to ten, or from zero to one hundred, and is typically further divided into smaller units (see **Figure 5.1**). These smaller units, which are exactly one tenth as long as the graduated scale, are further subdivided into fifths or tenths. The ocular micrometer is placed in the light path, exactly at the focal length of the ocular lens, where it will be in focus and clearly visible in the field of view; it will appear superimposed over the specimen. If the microscope is carefully focused and the ocular micrometer correctly positioned, both the specimen and the graduated scale will be in sharp focus.

THE STAGE MICROMETER

Unfortunately, the graduations on the ocular micrometer must be relative length measurements rather than exact ones (like millimeters or micrometers). This is necessary because it is not possible to make every objective lens exactly alike and because our microscopes usually have three or four objective lenses, each giving a different magnification. Remember, also, that the ocular micrometer is positioned at the focal length of the ocular lens and therefore is really being used to determine the dimensions of the image produced by the objective lenses. Consequently, the ocular micrometer must be calibrated for each objective lens, and those calibrations are only valid for the objective lenses and the micro-

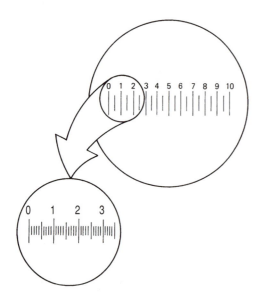

Figure 5.1   The ocular micrometer.

scope that they were determined in. The ocular micrometer is calibrated against a scale of known length (like a ruler) that is positioned on the microscope stage and focused like a specimen. This known scale is called a stage micrometer.

A typical *stage micrometer* (see **Figure 5.2**) has a graduated scale that is of known length. Stage micrometers with scales from 1.0 through 5.0 millimeters are available commercially. For illustrative purposes, we will assume that the scale is exactly one millimeter (1.00 mm), and that it is further subdivided into tenths ($1/10$; 0.10) and hundredths ($1/100$; 0.01). Calibration of the ocular micrometer requires that the image of the stage micrometer be superimposed on that of the ocular micrometer. Since the exact dimensions of the stage micrometer are known, the relative dimensions of the ocular micrometer can be easily calculated.

APPLICATIONS

Once an ocular micrometer has been calibrated, its scale can be used to determine the length and diameter of a cell (or any other object that can be examined under the microscope). As we will see in later exercises, this information can then be used to calculate cell volume and cell weight. If the total number of cells in the suspension is known, the total biomass of those cells can be calculated.

If the area of the field of view is known (calculated from the diameter), the number of cells in a known volume of fluid can be determined. A portion of the sample, typically 0.01 ml, is carefully layered over a known surface area and allowed to dry. The cells, or any other particles, can be stained and then counted. Relatively simple mathematics are used to determine the number of cells distributed over the surface, based on the average number of cells in each field of view.

LABORATORY OBJECTIVES

The use of the microscope as a quantitative tool requires an understanding of how the measurements are made and why these calibrations are necessary. You should:

- Understand the use of the ocular micrometer and how its relative units can be used to determine actual cell length and diameter.

- Understand how the number of cells observed in the field of view can be related to number of cells on the surface being examined or in the fluid spread over that surface.

MATERIALS NEEDED FOR THIS LABORATORY

This exercise may be performed as a demonstration, by groups of students, or individually as directed by your instructor.

1. **Microscopes equipped with ocular micrometers.** The ocular micrometer should be mounted at the focal plane according to the specifications provided by the manufacturer of the microscope.

Figure 5.2   The stage micrometer.

2. **Stage micrometers.** The stage micrometer should be carefully handled. Use only distilled water and lens paper to clean it handle it only by its edges. Keep the micrometer in its case at all times (except, of course, when you are using it) and be careful that nothing is allowed to scratch its surface.

LABORATORY PROCEDURE

A. DETERMINATION OF DIAMETER AND AREA OF FIELD OF VIEW

The field of view can be measured by focusing the microscope on the stage micrometer and directly measuring its diameter. Study the examples shown in **Figure 5.3.** Although not drawn to scale, the diagram clearly shows what you will observe when you focus each objective lens on the stage micrometer.

To improve accuracy, always use the same edge of the graduation mark as your reference line. As you increase the magnification, the width of the marks will appear to increase. You can avoid the error of including the thickness of the marks in your measurements by always using the same edge (right or left) of the marks.

1. Carefully align the stage micrometer on the microscope stage so that the graduated scale is approximately centered in the light path.

2. Center the low-power objective lens (10X) over the slide and carefully lower it until it is about ¹/₂ cm above the slide. If your microscope has the autostop feature, you should lower the lens until it stops.

3. Focus the image of the stage micrometer by moving the lens *away* from the stage. Alternately adjust the condenser and focus until a clear image is obtained.

4. Move the stage micrometer so that a convenient graduation is located against the left edge of the field of view.

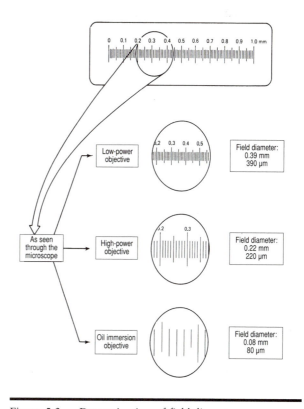

Figure 5.3   Determination of field diameter.

If your microscope does not have a flat field of view, you might need to adjust the focus so that the edges of the field of view are in sharp focus.

5. Determine the diameter of the field by observing the graduation closest to the right edge of the field. It might be necessary to extrapolate between graduations to determine the exact diameter.

6. Record the results in the appropriate blank on the Laboratory Report Form.

7. Repeat steps 3, 4, 5, and 6 for the other objective lenses on your microscope. Be particularly careful to ensure that there is no possibility of the lenses damaging the stage micrometer. Watch from the side when you change lenses. When focusing with the oil immersion lens, as always, use only the fine-adjustment.

B. CALIBRATION OF THE OCULAR MICROMETER

The ocular micrometer is calibrated by comparing its graduations with those on the stage micrometer. **Figure 5.4** shows, in idealized diagrams, how both micrometers might appear in a microscope field of view. In the example shown there are two points, indicated by arrows, where graduations on both micrometers coincide. When more than one such point is observed, use the longer distance for your calculations.

1. Ascertain that the ocular micrometer is correctly installed in the ocular lens. The best way to do this is to focus on a specimen. If an ocular micrometer is properly installed, you will be able to see it in focus with the specimen. *There is no reason to remove the ocular lens.* If you do not see the ocular micro-

meter, ask your instructor for assistance. If you are using a binocular microscope you will need to determine which lens contains the ocular micrometer. As you look through the instrument, close one eye at a time—the lens with the ocular micrometer will become apparent.

2. Center the stage micrometer on the microscope stage.

3. As you look through the low-power objective lens, focus the microscope until the graduations of the stage micrometer appear in sharp focus. Rotate the ocular lens until the two sets of graduations are parallel to each other (see **Figure 5.4**). It will be easier to complete the calibration if the two sets of graduations overlap somewhat.

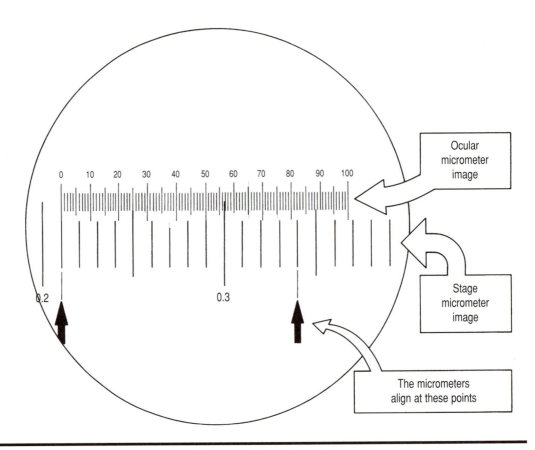

Figure 5.4   Calibration of the ocular micrometer.

4. Move the stage micrometer so that a convenient calibration mark is aligned with the zero mark on the ocular micrometer. Use the mechanical stage adjustments if you have them on your microscope.

5. Locate the point where another graduation on the ocular micrometer aligns with a graduation on the stage micrometer. In the example, there are two points of alignment, as indicated by the two arrows in the diagram. If more than one point of alignment is observed, use the one with the greatest length for your calculations. In this example, 82 units on the ocular micrometer align with 0.13 mm on the stage micrometer. (Note where zero is aligned.)

6. Record the alignment points for both micrometers in your notes.

7. Repeat steps 3, 4, 5, and 6 for each objective lens and record your data on the Laboratory Report Form.

8. Calculate the dimensions of the graduations on the ocular micrometer for each magnification. You can do this by dividing the length (in millimeters) observed on the stage micrometer by the number of units that coincide to that length on the ocular micrometer.

Ocular micrometer unit length =

$$\frac{\text{Stage micrometer length (in mm)}}{\text{Ocular micrometer units}}$$

For example, in Figure 3.4, you would divide 0.13 by 82.0, for a value of 0.0016. This indicates that each unit of the ocular micrometer is 0.0016 mm, or 1.6 μm, in length.

Ocular micrometer unit length =

$$\frac{\text{0.13 mm}}{\text{82.0 units}} = 0.0016 \text{ mm}$$

9. Record all data in the Laboratory Report Form. This completes the calibration of your microscope. You will need to refer back to these data in later exercises.

NOTES

Name _____

Section _____

LABORATORY REPORT FORM

EXERCISE 5
CALIBRATION OF THE MICROSCOPE

What was the purpose of this exercise?

A. RECORD THE FOLLOWING INFORMATION ABOUT YOUR MICROSCOPE:

1. Serial number of microscope (or identifying number): _____

2. Serial number of low-power lens: _____

3. Serial number of high-power lens: _____

4. Serial number of oil-immersion lens: _____

B. RECORD THE FOLLOWING DATA ABOUT THE FIELD OF VIEW
OF YOUR MICROSCOPE:

LENS	DIAMETER	AREA
Low power		
High power		
Oil immersion		

Units in millimeters (micrometers)

C. RECORD CALIBRATION DATA FOR THE OCULAR MICROMETER:

LENGTH OF ONE OCULAR MICROMETER UNIT

LENS	MILLIMETERS	MICROMETERS
Low power		
High power		
Oil immersion		

QUESTIONS:

1. Why must the ocular micrometer be located at the focal plane of the ocular lens?

2. Why isn't the actual diameter of the field of view observed at 1000 magnification to be exactly one tenth of that measured at 100 magnification?

3. If you determined that three spherical objects each had diameters equal to 1.65 ocular micrometer units at the three magnifications of your microscope, what would their actual diameters and area be?

LENS	DIAMETER	AREA
Low power		
High power		
Oil immersion		

Units should be in millimeters or micrometers

CELL DIMENSIONS:
MEASUREMENT OF CELL SIZE

The dimensions of a cell can be measured accurately if an ocular micrometer is installed in your microscope and if it has been correctly calibrated. In this exercise we will use the calibration data obtained in Exercise 5 to determine the dimensions of red blood cells and bacteria.

If the diameter and/or length of a cell is known, it is possible to calculate certain other biomass parameters, such as mean cell volume and mean cell surface area. These parameters have both clinical and ecological significance; if they can be accurately determined for individual cells, such data as total cell volume and total surface area for the population can also be calculated. Additionally, the weight of a cell can be estimated if the volume and density are known.

BACKGROUND

For the purposes of this exercise, we will assume cocci to be round, spherical bodies, not unlike tennis balls or basketballs. We must, however, realize that they are not truly round. Bacilli can be assumed to resemble cylinders whose length and diameter can be measured. Red blood cells, because of their concavity, are unique. Except for their diameter, the biomass parameters of red blood cells cannot be calculated easily.

The formulas needed to calculate the biomass parameters for cocci and bacilli are relatively simple. If cocci are treated as spheres, then the formulas used for circular and spherical bodies can be used. If bacil-li are being examined, the length of the bacillus and its circumference (calculated from diameter) can be used to determine surface area and volume.

The relationship between a cell's surface area and its volume is extremely important. The cell membrane, whose expanse is determined as surface area, is the only means by which a bacterial cell can absorb nutrients, exchange gasses, secrete enzymes, or excrete wastes. Without the rapid exchange of these materials, bacterial cells would be incapable of their rapid growth.

Before you begin this exercise, you should look up the formulas for the following and review their calculation:

1. Area of a circle:

2. Circumference of a circle:

3. Volume of a sphere:

4. Surface area of a sphere:

5. Surface area of a cylinder:

6. Volume of a cylinder:

LABORATORY OBJECTIVES

Much information about microbial populations, and about individual cells in those populations, can be derived from simple diameter and length measurements. In this exercise you should:

- Become familiar with procedures for measuring the length and diameter of cells.

- Understand the relationship between cell diameter and length and cell volume and surface area.

MATERIALS NEEDED FOR THIS LABORATORY

1. Microscopes equipped with ocular micrometers. If possible, use the microscope you calibrated in Exercise 5. If that is not possible, either obtain the data for the microscope you are using or calibrate the one you will use for this exercise.

2. Small test tubes containing about 5.0 ml of 0.9% saline

3. Sheep blood

4. Pasteur pipettes

5. Glass slides and coverslips

6. Twenty-four-hour broth cultures of:
   Bacillus cereus
   Staphylococcus aureus

LABORATORY PROCEDURE

A. DIAMETER OF RED BLOOD CELL

1. Prepare a suspension of red blood cells. Place several drops of sheep blood into

4. Repeat step 3. Use the next level of magnification to determine if a more accurate measurement might be obtained.

5. Calculate the mean cell diameter of the red blood cells.

6. Attempt to measure the thickness of red blood cells.

7. Record all you results in the Labboratory Report Form.

B. DIAMETER AND LENGTH OF BACTERIA

Bacterial cells are easier to see, and therefore easier to measure, if they are stained. However, the drying associated with the preparation of a smear almost always causes some distortion, usually shrinkage, of the cells. In this exercise, however, we will assume that there is only negligible distortion introduced by smear preparation and the simple staining procedures. As an alternate procedure, consider using wet mounts to measure the length and diameter of the cells.

Although bacilli can be viewed as cylinders, with the diameter of the cell usually remaining constant over its length (except for diphtheroids and ellipsoidal cells),

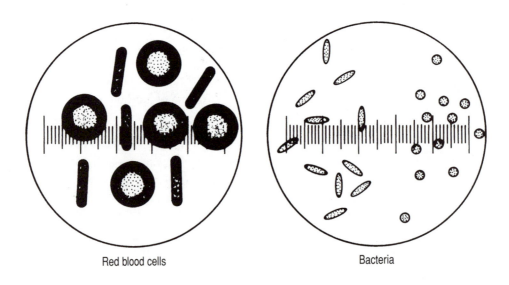

Red blood cells   Bacteria

Figure 6.1   Measurement of cell diameter (idealized drawing).

consider the length of a bacillus to be the total length, including any rounding of the ends (see **Figure 6.2**). This problem is not a serious one if we are only concerned with the length of the cell, but when we want to calculate biomass, we must consider how much error this assumption is likely to introduce.

1. Prepare simple stains (see Exercise 2) of the two bacterial suspensions. It is particularly important that you prepare smears with well-separated cells. If necessary, dilute the bacterial suspension with saline.

2. As an alternative procedure, add one loopful of crystal violet to 1.0 ml of a 24-hour broth culture, and use this preparation to make wet mounts.

3. Using the oil-immersion objective, determine the mean diameter of *Staphylococcus aureus* cells. You should take measurements of at least 10 cells to account for any variation in their diameter.

4. Similarly measure the length and diameter of *Bacillus cereus*. Again, measure at least 10 cells and calculate the mean. It is good practice to measure the length and diameter of the same cell and to keep the two measurements paired. Why?

5. Record all your results in the Laboratory Report Form.

C. CALCULATION OF CELL VOLUME AND SURFACE AREA

1. Calculate the surface area of the average *Staphylococcus aureus* cell. Use the

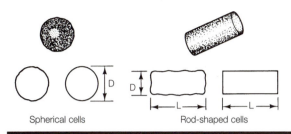

Figure 6.2 Dimensions of spherical and rod-shaped cells.

formula for the surface area of a sphere and assume π to be 3.14.

2. Calculate the volume of your average *Staphylococcus aureus* cell.

3. The calculation of the surface area of the average *Bacillus cereus* cell requires that you multiply the circumference of the cell by its length and then add the area of two ends. Use the diameter of the cell to calculate the area and circumference of a circle.

    a. Calculate the area of the wall of the cylinder (cell): Multiply the circumference by the length of the cell.

    b. Calculate the area of the end of the cylinder (cell): Calculate the area of the circle and multiply by two.

    c. Add the results of (a) and (b) together to obtain the surface area of the average *Bacillus cereus cell* in the suspension.

4. Calculate the volume of the rod-shaped cell by multiplying the area of the ends by the length of the cell.

NOTES

LABORATORY REPORT FORM

EXERCISE 6
CELL DIMENSIONS

What was the purpose of this exercise?

A. COMPLETE THE FOLLOWING TABLES:

	CELL MEASUREMENTS
Red blood cell diameter	
Thickness of red blood cells	
Diameter of stained *Staphylococcus aureus*	
Wet mount measurement	
Diameter of stained *Bacillus cereus*	
Wet mount measurement	
Length of stained *Bacillus cereus*	
Wet mount measurement	

CALCULATED VALUES FOR BACTERIAL CELLS

	Stained	Wet Mount
Staphylococcus aureus		
Surface area of average cell	——————	——————
Volume of average cell	——————	——————

CALCULATED VALUES FOR BACTERIAL CELLS (continued)

	Stained	Wet Mount
Bacillus cereus		
Surface area of average cell	_____	_____
Volume of average cell	_____	_____

QUESTIONS:

1. Did you experience any difficulty measuring the thickness of the red blood cells? If so, why do you think it was difficult?

SURVEY OF MICROORGANISMS

Microorganisms can be divided into two major groups dependent upon their cell structure—the *prokaryotes* and the *eukaryotes*. Prokaryotic organisms, those which lack a nucleus and other membrane bound internal organelles, would include the bacteria and cyanobacteria. Fungi, protozoa, algae, and the multicellular parasites would be examples of eukaryotic microorganisms. They are all characterized by having a distinct nucleus and membrane bound organelles. Viruses are not located in either of these groups as they are acellular organisms. Refer to your textbook for further descriptions of these organisms.

In this exercise, we will concentrate on examples of the cyanobacteria (prokaryotes) as well as the eukaryotic microorganisms.

BACKGROUND

Most of our laboratory exercises concentrate on the characterization and identification of bacteria and viruses—prokaryotic and acellular microorganisms. There are, however, a large variety of other organisms included in microbiology. These include the cyanobacteria, fungi, yeast, protozoa, algae and helminths.

The incidence of fungal and parasitic diseases in the United States has shown significant increase over the last several years. Contributing factors appear to be the increased numbers of immunocompromised individuals (those whose immune system is not capable of complete protection), environmental contamination, and our increasingly global society.

KINGDOM MONERA (PROKARYOTAE)

The organisms found in the Kingdom Monera (Prokaryotae), are bacteria and cyanobacteria, as shown in **Figure 7.1**. The cyanobacteria are photosynthetic, usually unicellular prokaryotic organisms. Some tend to form connected cells which appear as threadlike filaments. Because they are autotrophic (organisms which use carbon dioxide as their major carbon source), they do not pose a health threat to humans. They are, however, extremely important nitrogen-fixing organisms (capable of converting atmospheric nitrogen into ammonia which can be used by plants) and responsible for much of the oxygen produced world-wide.

KINGDOM PROTISTA

Eukaryotic unicellular or colonial organisms are in the Kingdom Protista (see **Figure 7.2**). Included are algae and protozoa; their close relationship can be seen in the dual classification often encountered, such as with the *Euglena*. What characteristics of the *Euglena* would cause its inclusion in the algae? In the protozoa?

ALGAE

Algae are photosynthetic eukaryotes characterized by the presence of membrane bound chloroplasts and the absence of true roots, stems, or leaves. Many of them exhibit motility by means of flagella or gliding. They are most commonly found in fresh water, with some marine and some soil species seen. Algae are significant primary producers (organisms which con-

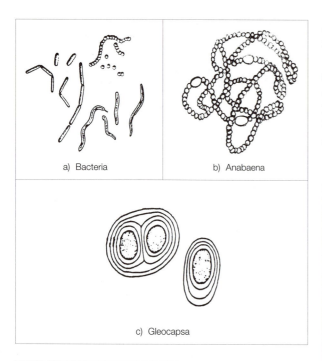

a) Bacteria   b) Anabaena

c) Gleocapsa

Figure 7.1   Representative members of the Kingdom Monera (Prokaryotae).

inorganic carbon as found in carbon dioxide into organic forms which can be used by other living things), forming the beginning of the food chain. The presence of algae in an aquatic environment can be indicative of its overall quality. When excessive nutrients become present in a fresh water system, it may result in an algal "bloom" or the presence of excessive numbers of algae.

The microscopic algae, or subkingdom Algae, are often divided into three sub-groups—the Dinoflagellata or dinoflagellates (pigmented organisms which are motile by flagella), the Bacillariophyta or diatoms (unicellular algae with a silica cell wall), and the Chlorophyta or green algae.

PROTOZOA

Protozoa are unicellular or colonial, heterotrophic (requiring an organic carbon source) eukaryotes which lack a cell wall. They are usually found in moist and wet areas, and may be either free-living or parasitic. Classification of the protozoa has traditionally been based on their method of motility—a readily observed trait that may or may not have any evolutionary significance. The groups we will be looking at include the *Sarcomastigophora,* the

Apicomplexa and the *Ciliophora.* Human parasitic forms are seen in each of these phyla.

The *Sarcomastigophora* include the sub-phyla *Sarcodina* and the subphyla *Mastigophora.* The *Sarcodina,* or those protozoa that move by pseudopodia, include the parasite *Entamœba histolytica* which causes amœbic dysentery. The *Mastigophora,* or flagellated protozoa, include the

PROTOZOA

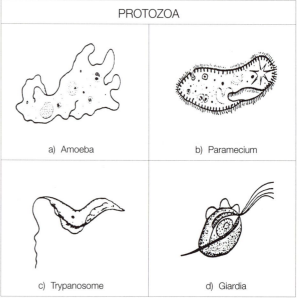

a) Amoeba   b) Paramecium

c) Trypanosome   d) Giardia

ALGAE

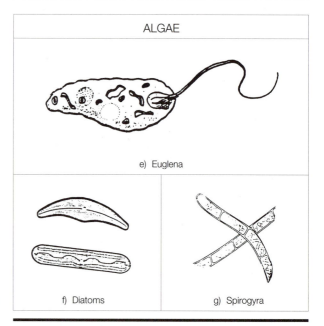

e) Euglena

f) Diatoms   g) Spirogyra

Figure 7.2   Representative members of the Kingdom Protista.

parasites *Trichomonas vaginalis* (a sexually transmitted organism), *Giardia lamblia* (the organism largely responsible for protozoan gastrœnteritis in day care centers), and *Trypanosoma gambiense* (causative agent of African sleeping sickness).

The *Apicomplexa* contain primarily non-motile protozoa, referred to as the sporozoans. These organisms are often characterized by complex life cycles involving more than one host and are always parasitic. Examples would include the various species of *Plasmodium* (causative agents of malaria), *Toxoplasma gondii* (a cat parasite that can cause severe congenital defects in humans), and *Cryptosporidium* (responsible for outbreaks of water-borne gastroenteritis) .

The ciliated protozoa belong to the group *Ciliophora*. The only human parasite that belongs to this group is *Balantidium coli* which causes diarrhea.

KINGDOM MYCETAE

Included in the Kingdom Mycetae are both the molds and yeasts. They are unicellular or multicellular, non-photosynthetic eukaryotes, which are surrounded by a chitin or cellulose cell wall and are highly adapted to a variety of environments.

MOLD

The molds are multicellular filamentous fungi composed of individual strands or hyphae organized in a branched meshwork known as the mycelium. They reproduce through the production of asexual and/or sexual spores. The classification of the fungi is based largely upon the type of spore production (see **Figure 7.3**). The saprophytic molds are important decomposers as they obtain their nutrients from dead organic matter or organic waste. They are also rarely parasitic on healthy plants or animals, instead serving as opportunistic pathogens (an organism which is not usually cause disease but can become pathogenic under special circumstances).

We will be looking at examples of the *Ascomycota* (*Aspergillus* and *Penicillium*) and *Zygomycota* (*Rhizopus*).

YEAST

The yeasts are nonfilamentous, unicellular fungi which reproduce by budding. Their identification is based upon both morphology and biochemical tests. Yeast are commonly isolated on the surface of plants or amongst the commensal organisms comprising

our normal flora (a relationship between two species where the one species benefits and the other is neither benefitted nor harmed). The yeast are facultative anaerobic organisms, allowing them to grow both in the presence of oxygen, where they undergo aerobic cellular respiration, and in its absence, where they undergo fermentation. This anaerobic fermentation results in the production of alcohol and carbon dioxide and is the basis of the leavening of bread dough and the production of alcoholic beverages.

Yeast are excellent examples of opportunistic organisms which do not normally cause disease, but can when there is immunosuppression due to illness (AIDS), destruction of other normal flora (antibiotic treatment), or chemical changes in the body (alteration in vaginal secretions with pregnancy). Yeast infections in the forms of thrush or vuvlovaginal candidiasis are not uncommon under these circumstances.

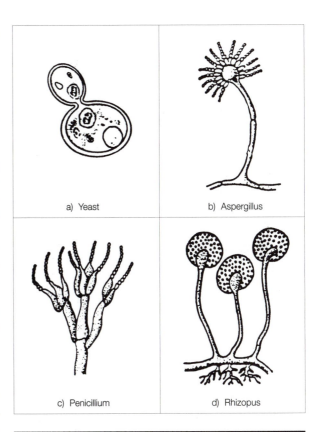

Figure 7.3   Representative members of the Kingdom Mycetae.

KINGDOM ANIMALIA—HELMINTHS

The helminths are multicellular eukaryotic organisms which include several phyla of medical and veterinary significance (see **Figure 7.4**). The Platyhelminthes (flatworms) include the fluke and tapeworms. Many of these intestinal parasites have complex life cycles which involve not only man but polluted water, and intermediate hosts such as fish or snails. The Nematodes (roundworms) include the pinworms and hookworms. Many of the roundworms have much simpler life cycles and are present in large numbers in soil, fresh water and seawater.

Helminths are included in microbiology as their identification is usually done by microscopic examination of stool specimen where one looks for the eggs, or body fluids and biopsy tissue which are examined for the presence of the parasite itself.

LABORATORY OBJECTIVES

- Distinguish, by kingdom, the eukaryotes observed.

- Name three opportunistic pathogenic molds, the medical problems that they cause in humans, and two beneficial molds.

- Name one pathogenic yeast and the medical problems it causes in humans.

- Identify the distinctive characteristics of each group of protozoans studied.

- Characterize helminths and discuss the reason for their inclusion in microbiology.

MATERIALS NEEDED FOR THIS LABORATORY

Your instructor may have you perform all or only select portions of this exercise. In addition, demonstration slides may be substituted for the living cultures.

A. KINGDOM MONERA (PROKARYOTAE)

    1. Living culture—*Gleocapsa*

    2. Prepared slides
       Gleocapsa
       Anabaena

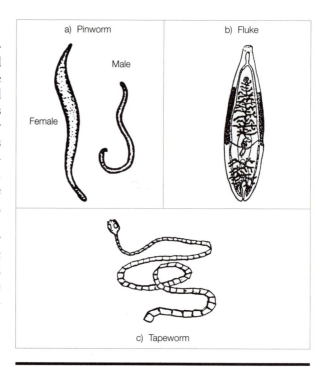

Figure 7.4   Representative members of the Kingdom Animalia (Helminths).

B. KINGDOM PROTISTA

    1. Living cultures
       Euglena
       Amœba
       Paramecium

    2. Prepared Slides
       Euglena
       Giardia
       Amœba
       Plasmodium
       Paramecium
       Trypanosoma
       Algae, mixed

    3. Microscope Slides

    4. Coverslips

    5. Pasteur pipettes

    6. Methylcellulose

    7. Toothpicks

C. KINGDOM MYCETAE

1. Sabouraud agar cultures and/or prepared slides
    Aspergillus
    Rhizopus
    Penicillium

2. *Saccharomyces cerevisiae* culture or yeast suspension

3. Prepared slide—*Pneumocystis carinii*

4. Microscope Slides

5. Coverslips

6. Pasteur pipettes

D. KINGDOM ANIMALIA

1. Preserved specimen or prepared slides
    Schistosoma
    Trichinella
    Clonorchis
    Ascaris
    Taenia
    Enterobius

E. MIXED LIVING ORGANISMS

1. Pond Water

2. Hay Infusion (1 week)

3. Microscope Slides

4. Coverslips

5. Pasteur pipettes

6. Methylcellulose

LABORATORY PROCEDURE

A. KINGDOM MONERA (PROKARYOTAE)

1. Prepare a wet mount of *Gleocapsa* (see Exercise 2).

2. Observe under high power. Record your observations on the Laboratory Report Form, Part A.

3. Observe the prepared slide. How does it compare to the live specimen? Record on the Laboratory Report Form, Part A.

B. KINGDOM PROTISTA

1. Prepare wet mounts of the protozoan cultures.

2. Observe wet mounts using high power. If the organisms are moving too fast, add a drop of methylcellulose to your next drop and mix with a toothpick before adding the coverslip. Record your observations on the Laboratory Report Form, Part B.

3. Observe the prepared protozoan and algae slides. Record on the Laboratory Report Form, Part B.

C. KINGDOM MYCETAE

1. If Sabouraud agar cultures are available, tease a small amount of growth with a dissecting or inoculating needle. Place in a drop of water in the center of your slide, and prepare a wet mount.

2. Observe slides under high power. Record your observations on the Laboratory Result Form, Part C.

3. Observe prepared slides of the fungal specimen. Compare your observations with those of living specimen. Record on the Laboratory Report Form, Part C.

4. Place a drop of yeast suspension on a clean slide. Add a drop of methylene blue stain. Prepare a wet mount. Observe under high power and record on the Laboratory Report Form, Part C.

5. Observe a prepared slide of *Pneumocystis carinii*. Record your results on the Laboratory Report Form, Part C.

D. KINGDOM ANIMALIA

1. Observe the prepared slides and preserved specimen of the helminths. Record your results on the Laboratory Report Form, Part D.

E. MIXED LIVING ORGANISMS

1. Prepare wet mounts of the pond water and/or hay infusion samples. Observe them under high power.

2. Using resources available, attempt to identify the organisms viewed. If organisms are moving too fast to view

details, you may add a drop of methyl-cellulose to your next drop.

3. Share your observations with your laboratory partners.

4. Record your results on the Laboratory Report Form, Part E.

Name _____

Section _____

LABORATORY REPORT FORM

EXERCISE 7
SURVEY OF MICROORGANISMS

What was the purpose of this exercise?

A. KINGDOM MONERA (PROKARYOTAE)

1. Draw representative samples of the Gleocapsa as viewed in a wet mount.

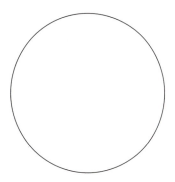

How does the prepared specimen compare to the live specimen?

B. KINGDOM PROTISTA

Record your observations of the varied protozoan and algal specimen.

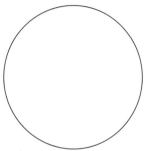

Organism _____
Magnification _____

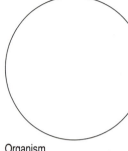

Organism _____
Magnification _____

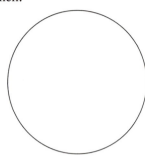

Organism _____
Magnification _____

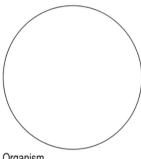

Organism _____
Magnification _____

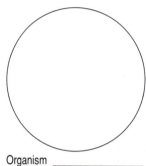

Organism _____
Magnification _____

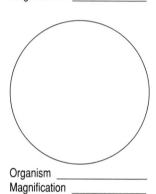

Organism _____
Magnification _____

C. KINGDOM MYCETAE

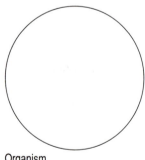

Organism _____
Magnification _____

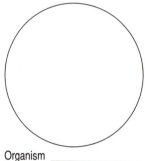

Organism _____
Magnification _____

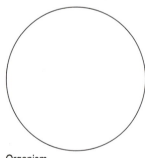

Organism _____
Magnification _____

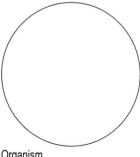

Organism _____
Magnification _____

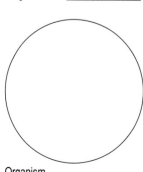

Organism _____
Magnification _____

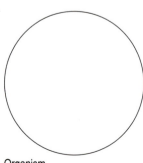

Organism _____
Magnification _____

D. KINGDOM ANIMALIA

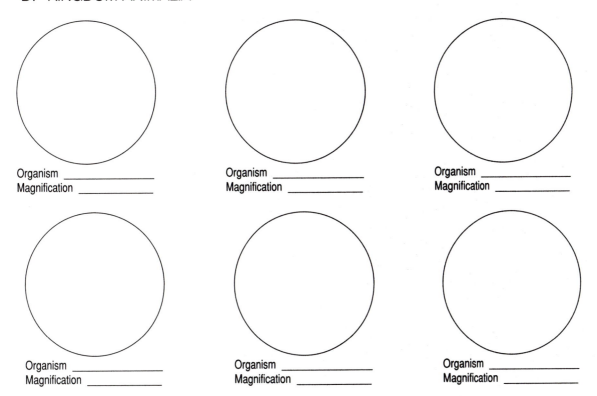

Organism _____
Magnification _____

Organism _____
Magnification _____

Organism _____
Magnification _____

Organism _____
Magnification _____

Organism _____
Magnification _____

Organism _____
Magnification _____

E. MIXED LIVING ORGANISMS

Draw any protozoans found in the pond water anf describe their method(s) of motility.

QUESTIONS:

1. How do the cyanobacteria compare to the bacteria you have previously observed?

2. What similarities did you observe between the protozoa and the algæ? What differences?

3. Algal blooms often occur in summer months. What factors would contribute to their occurrence?

4. Compare mold spores with bacterial endospores.

5. Give an example of an opportunistic fungi. Are all medically important fungi opportunistic?

6. What are the significant ingredients of Sabouraud's agar? What makes it more selective for fungi than bacteria?

7. Diagram the life cycle of *Schistosoma mansoni*. Propose a public health control plan for the elimination of this organism.

ISOLATION TECHNIQUES

Early in the study of microbiology, Robert Koch recognized that in order to identify the causative agent of disease, it was essential for organisms to be separated from one another and grown in a pure culture. The developments of these techniques formed the basis for his Postulates which are still utilized today.

The procedures that are used in microbiology to accomplish the separation of bacteria into discrete types are referred to as *isolation procedures.* This enables the isolation of bacteria into pure cultures. A *pure culture* is a culture consisting of a single type (species or strain) of bacteria, usually derived from a single cell. Single cells grow into a mass of cells large enough to see. Such a mass of cells is referred to as a *colony.*

Suspensions of cells are spread over the surface of, or within, a nutrient medium in such a way that each cell will grow into a colony that is physically separated from any other colony. Since colonies are assumed to be clones, isolated colonies represent pure cultures.

In this exercise, we will attempt to isolate pure cultures of bacteria from mixtures. We will also examine the appearance of the colonies and determine the colonial morphology of the organisms studied. It is important to remember that when we isolate a pure culture from a mixed culture, the colonies that grow may contain a bacterium that the microbiologist is seeing for the first time. The recognition of colony types on the basis of their morphology is the critical first step in diagnostic bacteriology; it is the basis of all clinical isolations.

BACKGROUND

If we are going to be able to grow bacteria, we must provide them, under suitable environmental conditions, with all the nutrients they need. The mixture in which the nutrients are supplied is referred to as the *growth medium* (plural: *media*). The medium also provides the necessary moisture and controls the pH of the environment.

Some media are solidified with a colloidal polysaccharide derived from seaweed (red algae), called *agar.* Agar has no nutrient value and cannot by hydrolyzed (broken down to low molecular weight compounds) by most commonly cultured bacteria. It is simply added to the liquid components to solidify the medium. If the medium is used in the liquid form (without the addition of agar), it is called a *broth.* The use of the word *agar* in the name of a medium means that it is a *solid;* the use of the word *broth* means it is a *liquid* medium.

AGAR

Agar, as we have seen, cannot be hydrolyzed by most bacteria and is of no nutrient value to the bacteria. The significance of this is that it can be used as a solidifying agent for bacterial media without changing the components of the medium or without liquefying during the time the bacteria are growing. Agar has several other properties that are also of great importance to the microbiologist, including:

1. **Remains solid at growth temperatures.** Agar was discovered as the result of a search for a means of

separating bacteria into pure cultures. Gelatin, on the other hand, "melts" when the temperature is raised above 25°C and is readily hydrolyzed by many bacteria. When gelatin is used as a solidifying agent, as soon as it liquefies (either because of "melting" or because of bacterial activity) the bacteria become mixed together and can no longer be studied as pure culture populations. Agar does not liquefy until the temperature is raised above about 100°C, and, since it is usually not liquefied by the action of bacteria, can be used to solidify culture media without having to be concerned about the mixing of pure cultures.

2. **Agar is relatively transparent.** Solidified agar is slightly cloudy, but still somewhat transparent. This is an important attribute because it is possible to observe many characteristics of bacterial colonies that would not be apparent if the solidifying agent were opaque.

3. **The physical properties of agar.** Agar is a reversible colloid that can change from the sol (liquefied) phase to the gel (solidified) phase at certain temperatures. The gel to sol (solid to liquid) phase change occurs at about 100°C, but the reverse change, from sol to gel (liquid to solid) occurs at about 45°C. This means that agar can be heated to its liquefaction temperature and then cooled to about 55°C and held in the liquid state until used. While it is liquid, additional components can be added to it and, when necessary, bacteria can be mixed in the agar for separation. If some heat-labile nutrients or bacteria are added to liquid agar when it is at 50°C, the agar can be dispensed into tubes or plates without heat-denaturation of the nutrients or the heat-killing of the bacteria. This, of course, would not be possible if the medium solidified at a higher temperature.

The Use of Agar

When agar solidifies, it retains its shape until it is liquefied again. There are several ways that solidified agar is used (see **Figure 8.1**):

A *slant* is a culture tube (or test tube) in which the medium has solidified while the tube was held in a slanted position. The culture tube is filled about 1/3 full with liquefied medium which is allowed to solidify so that the medium has a substantial slant. The slant provides a good growth surface and is an ideal way of maintaining cultures for study.

A *stab,* unlike a slant, is made by allowing the medium to solidify while the culture tube is held in an upright position. The tube is filled between 1/3 and 1/2 full. Stabs are used when reduced oxygen is needed or when it is desirable to observe the effects of growth of bacteria *in* the medium rather than *on* the medium.

A *deep* or *pour* usually contains between 18 and 20 ml of agar medium. They are usually used to make *pour plates.* Tubes of media are considerably easier to store than are plates, making them easier to handle and manipulate. The medium is liquefied and held in a water bath at about 50-55°C. Components and/or bacteria can be added to it as experimental design requires.

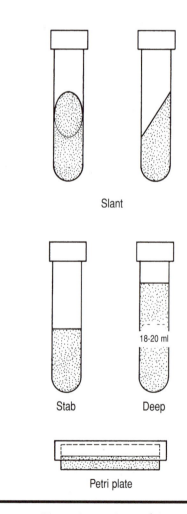

Figure 8.1   Uses of Agar

A *petri plate* is a glass or plastic dish with an overlapping cover. Melted medium is poured into the dish and allowed to solidify. The resulting surface of the medium is used to culture bacteria and to separate them into pure cultures by a procedure known as the *streak plate.* In the streak plate the bacteria are spread *over the surface* of the medium, while in the pour plate, the bacteria are *submerged in* the medium.

ISOLATION PROCEDURES

There are two isolation procedures commonly used in microbiology. The first procedure, the *streak plate,* is probably the single most-used procedure in any microbiology laboratory. The mixture of bacteria is spread over the surface of a solid nutrient medium so that isolated colonies develop. The second procedure, the *pour plate,* is less commonly used for routine isolation, but is frequently used to estimate the numbers of bacteria in a sample. It requires that a portion of the sample be mixed with the melted medium, which is then poured into a petri plate and allowed to solidify. If the sample was properly diluted, isolated colonies will develop after incubation.

Streak-Plate Procedures

The successful streak plate will have isolated colonies that can be transferred to fresh media with confidence that the cultures are pure.

1. A small amount of colony material is picked form the source culture with a sterile inoculating loop. If the source culture is liquid, a sterile loop is simply dipped into the sample, using proper aseptic technique.

2. The needle or loop is then streaked over the surface of a nutrient medium in such a way as to spread the sample over the medium. As the needle or loop moves across the surface of the medium, most of the bacteria are deposited at the beginning of the streak. Toward the end, however, the remaining bacteria are placed on the medium so that when they grow, isolated colonies will develop (see **Figure 8.2**). It is important that you maintain careful aseptic technique throughout this process or you will achieve nicely isolated contaminants.

3. Failure to obtain isolated colonies can almost always be traced to one of two common mistakes:

   a. Using *too much inoculum:* If you transfer too much sample to the streak plate, there will not be sufficient surface area to spread out the bacteria. A good rule of thumb is that if you can see more than just a speck of culture material on the end of the needle, you have used too much.

   b. Using *too little surface area:* A single wavy line down the center of the plate will not produce isolated colonies. *Use as much of the surface of the medium as possible.* Place the streaks as close together as possible without overlapping them. The pattern you choose is not nearly as important as whether or not you get isolated colonies.

4. The needle and loop will dig into the media if too much pressure is applied when making the streak. There should be only enough pressure to "steer" the instrument; rely on its weight and balance to provide most of the downward pressure. Practice on the countertop (you do not need to flame the needle this time). If you do cut into the medium (and only practice will make it unlikely that you will), stop the streak and carefully remove the needle or loop to minimize splattering when the tension on the instrument is released.

Pour-Plate Procedures (see **Figure 8.3**)

The pour plate differs from the streak plate in that the sample is mixed with melted medium, which is then poured into a petri dish. After the medium solidifies, the bacteria are held in place by the solidified medium; isolated colonies develop after appropriate incubation.

1. An aliquot (portion) of a liquid sample is transferred with a sterile pipette to liquid medium that is held in a water bath at about 50°C. The medium and sample are then gently mixed (gently enough so that *no bubbles* are formed) and poured into a petri dish.

a) Steps for a three-section streak plate.

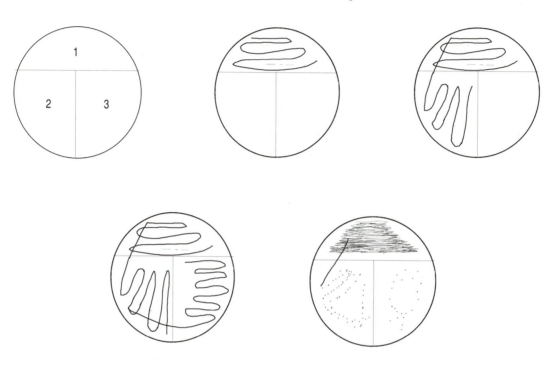

b) Alternative streak patterns

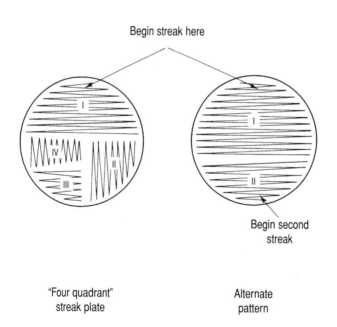

Begin streak here

Begin second
streak

"Four quadrant"
streak plate

Alternate
pattern

Figure 8.2   Preparation of a streak plate.

After the medium solidifies, it can be incubated and the isolated colonies observed and selected.

2. If the pour plates are to be used for counting of bacteria, they should have between 30 and 300 colonies on them for the counts to be considered valid. More than 300 colonies usually means an overcrowded plate where some colonies are too small to be seen or did not grow at all. What is the reason for the lower limit of 30?

3. One of the most common problems encountered by beginning microbiologists is that of determining the correct dilution needed to produce isolated colonies. Experience is the best teacher, but a good rule is to try to use dilutions that will bracket the correct one. For example, most experienced microbiologists will use three or four dilutions just to be sure of getting plates that will have between 30 and 300 colonies.

4. The simplest way of diluting your sample is to transfer a small amount (typically 1.0 ml) of the sample into a known volume of either saline (at 0.9% NaCl) or water. If 9.0 ml of diluent (saline or water) is used, you will have made a 1:10 dilution, and if you use 99 ml, a 1:100 dilution will result. These dilutions steps can be combined in any order to dilute the sample as much as needed. In this exercise, we will make several 1:10 dilutions of the sample and transfer one loopful of the diluted samples into the medium.

QUANTITATIVE ESTIMATIONS

If you know the volume of the original sample used to make the first dilution for a pour plate, and the final dilution used, you can easily determine the number of bacteria present in the sample. You simply have to multiply the three variables together to obtain the answer. Most microbiologists will report the result as the number of "bacteria/ml" or "colony-forming units (CFU)/ml." For example, the results of a urine culture might be reported as "9,750 bacteria/ml" or as >100,000 bacteria/ml. This method of quantification is commonly referred to as *plate counts*.

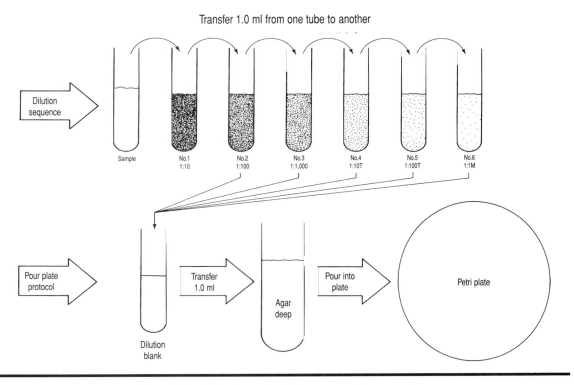

Figure 8.3   Dilution and plating protocol for pour plates.

The streak plate can be used to obtain a very rough approximation of the number of bacteria in the original sample. If the plate and streak pattern is mentally divided into four quadrants, the appearance of colonies in each quadrant can be used as a relative measure of the numbers present in the original sample. The most common system used is a 4+, 3+, 2+, and 1+ system, indicating in which quadrants the colonies were observed. For example, a 3+ would indicate that colonies appeared in the first three quadrants, but not in the fourth. This type of estimation is almost universally used by clinical laboratories. For example, the laboratory report will read "3+ alpha-hemolytic *Streptococcus*" or "1+ beta-hemolytic *Streptococcus*," and so on.

This application, as well as the use of the calibrated-loop streak plate (discussed in Exercise 30), are commonly referred to as the **quantitative streak plate**. We will not quantify our results at this time, but you will in a later Exercise.

COLONIAL MORPHOLOGY

Bacterial colonial morphology refers to the physical appearance of isolated colonies. Remember, when a colony is grown on an isolation plate, you will often be observing any physical characteristics of the bacterial isolate for the first time. These physical characteristics are often specific for the type of bacterium making the colony and can be used as a means of recognition. Colony morphology is, however, influenced by the medium and other growth conditions. The colonial morphology of the same bacteria may vary on different media or under differing environmental conditions.

Note: Colonial morphology is usually determined on colonies growing on the surface only. Colonies embedded in the medium, as with a pour plate, are surrounded by the medium and are limited by the physical characteristics of the medium. Colonies submerged in medium tend to be lenticular, or football-shaped, because the colony growth tends to split the medium, allowing growth only in the "bulge" produced.

How to Recognize Colonies

Colony recognition is an individual accomplishment. We can show you where to look for differences, but you must recognize those differences. For example, if everyone in your class was asked to write a descrip-

tion of your laboratory instructor, would you all come up with identical descriptions? Even though your descriptions are different, were any of them wrong? Did any of them fail to accomplish the objective of picking out the laboratory instructor from a group? We will suggest a list of characteristics that you can use to identify colony types. Whether you use them all or not is up to you—so long as you are able to recognize that colonies do differ and then are able to pick them out of a mixed group of colonies.

Any characteristic that allows you to differentiate and recognize colony types may be used. The list below is meant to be a guide—use whichever characteristics are necessary for you, not your partner, to recognize the colony types.

Colony Physical Characteristics (See **Figure 8.4**)

1. Colony shape and appearance:
Convex	Concave
Flat	Center plateau
Dimpled	"Fried-egg" shape

2. Colony consistency
Smooth	Mucoid
Dry	Rough
Fibrous	Granular

3. Colony edge:
Spreading	Smooth
Rhizoid	Lobate

4. Colony color:
   Color of the colony
   Color of the medium
   Iridescence

5. Changes in medium surface:
   Some bacteria dissolve agar and will appear to sink into it.

LABORATORY OBJECTIVES

You will be required to make several streak plates and at least one set of pour plates. The streak plates will be made from pure cultures and from a mixture containing those same bacteria. After incubation, you will determine the colonial morphology of the pure cultures and then attempt to find them in the mixed culture.

The pour plates will be made from either the pure cultures (if they are in broth) or from the mixture. You should compare the colonies produced by submerged bacteria with those growing on the surface.

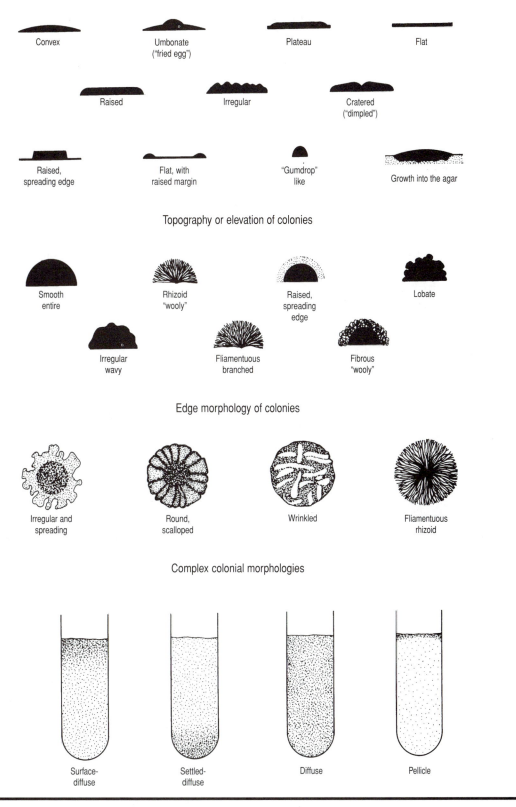

Figure 8.4   Some cultural characteristics of bacteria.

In this exercise you will learn how to isolate mixtures of bacteria into pure cultures and to recognize the specific colony morphology of some bacteria. You should:

- Understand the principles of bacterial isolation.

- Be able to explain the streak plate and the pour plate and the differences between the two.

- Be able to explain how each can be used quantitatively.

- Be able to write a description of some bacterial colonies.

- Be able to identify colonial types in a mixture.

MATERIALS NEEDED FOR THIS LABORATORY

A. STREAK PLATES

1. Nutrient agar plates—1 per pure culture + 1 per mixed culture available. The plates should have been prepared 24 hours ahead of time and incubated. This will allow the plates to dry so that there is no surface moisture to ruin your streak plates (especially with motile bacteria). It is also a good quality-control practice to eliminate plates that were accidentally contaminated when they were prepared.

2. Twenty-four hour agar slant cultures of:
    Staphylococcus epidermidis
    Pseudomonas aeruginosa
    Serratia marcescens (pigmented strain)
    Bacillus subtilis

3. Twenty-four hour mixed broth cultures of the following:
    Staphylococcus epidermidis
    Pseudomonas aeruginosa
    Serratia marcescens

B. POUR PLATES

1. *Five* nutrient agar deeps. These will need to be melted and then held at 50°C to keep them liquid.

2. Twenty-four hour mixed broth culture of the following:

Staphylococcus epidermidis
Pseudomonas aeruginosa
Serratia marcescens

3. Five sterile petri dishes

4. Six water-dilution blanks containing 9.0 ml of sterile water

5. Sterile pipettes

LABORATORY PROCEDURE

A. STREAK PLATES

1. Obtain and label the nutrient agar plates you will need for the streak plates. You will need one for each pure culture and one for each mixed culture.

2. Using aseptic techniques as demonstrated by the instructor, make streak plates of all the cultures. Remember that you should not use too large an inoculum and that you should use as much of the surface of the medium as possible. Avoid digging into the medium. If you make a mistake, repeat the streak plate with a fresh plate.

3. Place all the plates in a 37°C incubator in an inverted position (why?). Allow them to incubate for 24 – 48 hours.

4. Following incubation, record your observations. Write a description of each colony type and try to identify each type in the mixture.

5. If you did not get isolated colonies, examine your streak for possible reasons why. Repeat the experiment until you have obtained isolated colonies if instructed.

6. Record your observations on the Laboratory Report Form, Part A.

B. POUR PLATES

1. Obtain and label six dilution blanks and five sterile petri dishes. Obtain the necessary sterile pipettes. *Do not open the petri dishes.*

2. Obtain five tubes of nutrient agar deeps. Melt them in a boiling water

bath. Then maintain them in hot water bath to keep the tubes liquefied at a temperature of approximately 50–55°C. If the tubes are melted for you, *do not remove them from the water bath until you are ready to use them.* If your medium solidifies in the test tube, you must start over. You should have everything set up before you obtain the agar deeps. Once they are removed from the water bath they will solidify rapidly. *Examine* **Figure 8.3** *as you read the instructions.*

3. Using aseptic techniques, transfer 1.0 ml of your sample to the first dilution blank. Mix well. You now have a 1:10 dilution.

4. Transfer 1.0 ml from the first dilution blank to the second dilution blank. Mix it by shaking, and then transfer 1 ml to the next dilution blank. Continue in this manner until all of the dilution blanks have been innoculated, transferring 1 ml from each dilution tube into the subsequent dilution tube.

5. From this point on, you will be making pour plates of each dilution. *Now is the time to ensure that the medium is ready and petri dishes are in place.*

Dilution Blank No.	Agar Deep No.	Petri Plate No.
1	Not plated	Not plated
2	1	1
3	2	2
4	3	3
5	4	4
6	5	5

6. After mixing dilution tube #2, transfer one loopful of bacterial suspension to agar tube #1. Gently mix the medium and immediately pour it into a petri dish. Allow the petri dish to remain undisturbed while you continue.

7. Repeat the above until all the tubes melted agar and petri dishes have been used. You should have five plates by the time you are finished.

8. After the medium has solidified, place all plates in a 37°C incubator. Allow them to incubate for 24 – 48 hours.

9. Record your observations in the Laboratory Report Form, Part B. Write a description of each colony type and try to identify each type in the mixture.

10. Compare the colonies obtained with the pour plates with those obtained on the streak plates.

11. Record your results in the Laboratory Report Form, Part B.

NOTES

LABORATORY REPORT FORM

EXERCISE 8
ISOLATION TECHNIQUES

What was the purpose of this exercise?

A. STREAK PLATE

1. Describe the isolated colonies observed on the streak plates made from the pure cultures.

ORGANISM	Staphylococcus epidermidis	Pseudomonas aeruginosa	Serratia marcescens	Bacillus subtilis
COLONY APPEARANCE				
COLONY TOPOGRAPHY				
COLONY CONSISTENCY				
COLONY EDGE				
COLONY COLOR				
CHANGES IN MEDIUM				
AMOUNT OF GROWTH (4+, 3+, 2+, 1+)				

2. Sketch the appearance of your streak plate from the mixed culture.

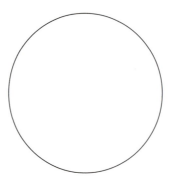

B. POUR PLATE

1. What dilution of original culture is present in each of the following dilution blanks?

DILUTION BLANK	DILUTION
#1	1:10
#2	
#3	
#4	
#5	
#6	

2. Complete the following table. If the plate contains between 30 and 300 colonies per plate, indicate the exact count. If there are more than 300 colonies per plate, indicate TNTC (too numerous to count). If there are less than 30 colonies per plate, indicate TFTC (too few to count).

PETRI PLATE #	DILUTION	# COLONIES/PLATE
1		
2		
3		
4		
5		

3. Which dilutions gave you isolated colonies?

4. Compare the results on the streak plate with those in your pour plates.

QUESTIONS:

1. How could the pour plate procedure be used for counting the number of bacteria in a sample?

2. Why do the submerged colonies look different than the ones growing on the surface of the medium?

3. Explain a laboratory report of a urine culture that reports "2+ *E. coli.*"

4. How would you know if a bacterial culture produced a water-soluble pigment?

5. How would you know if a bacterial culture produced a pigment, but did not excrete it?

6. How could you determine if contaminants were present on one (or more) of your plates?

7. Is it possible for organisms to grow on a streak plate, but not in the pour plate? Explain.

8. What advantages are there to the streak plate? the pour plate?

EXERCISE 9

OXYGEN REQUIREMENTS
OF MICROORGANISMS

A significant number of pathogenic bacteria are anaerobic. In addition to the well-known species of *Clostridium* (*C. tetani, C. botulinum, C. histolyticum,* etc.), there are many anaerobic and microaerophilic bacilli and cocci that are frequently encountered in clinical samples.

Cultures of the peritoneal cavity and of wounds are perhaps the most likely sources of anaerobic bacteria, although as anaerobic techniques have become more and more reliable, anaerobic bacteria are increasingly being found to be involved in pathogenic conditions. The culture and isolation of anaerobic bacteria involve enrichment in a semi-fluid medium followed by streak-plate isolation under anaerobic conditions.

BACKGROUND

Bacteria may be classified into four groups according to their requirements for molecular oxygen. In this exercise, these categories will be used to describe the species of bacteria being studied and not to describe a particular type of metabolism or reaction. As we will see, bacterial species that use anaerobic metabolic pathways may be called aerobes if they grow in the presence of oxygen (even though they may not use the oxygen). These groups are:

1. An *aerobic* bacterial species is able to grow in the presence of molecular oxygen. If the species *requires* molecular oxygen for metabolic energy production, it may be referred to as an *oblig-ate aerobe;* it cannot grow unless molecular oxygen is present in the environment. Such bacteria usually do not ferment

sugars, but oxidize them by a respiratory pathway. If on the other hand, the bacterial species does not require oxygen, and is not inhibited by it, it will be able to grow whether or not oxygen is present.

2. An *anaerobic* bacterial species is one that cannot grow in the presence of free (molecular) oxygen. Oxygen is toxic to these bacteria; they will only grow in environments that are free of oxygen. In some instances, very small amounts of oxygen are toxic, and the requirement for an oxygen-free environment is absolute. Such bacterial species are often referred to as *obligate anaerobes.*

3. Bacteria that require small amounts of oxygen are referred to as *microaerophilic;* they grow best in an environment that has about 5% oxygen. They also prefer an elevated carbon dioxide atmosphere. In a way, the difference between the *obligate anaerobe* and the *microaerophilic* is one of degree. Both groups may be considered to be sensitive to oxygen, but one is much more so than the other.

4. Bacteria that are able to grow without regard to the oxygen content of the environment are referred to as *facultative* organisms. They are not inhibited by oxygen, nor do they require it for their metabolism. There are at least two conditions that make an organism facultative:

a. Bacteria that are able to use either an aerobic (aerobic respiration) or anaerobic (anaerobic respiration and/or fermentation) energy-producing metabolism. These bacteria switch from one pathway to another, depending

79

upon whether or not oxygen is present in the environment.

b. Bacteria that use anaerobic metabolic (fermentation) pathways, but which are not inhibited by oxygen. These bacteria will grow in the presence of oxygen, but do not use it. They are sometimes referred to as *aerotolerant* organisms.

For the purposes of this exercise, species of bacteria may be defined according to the following:

Aerobic: A bacterial species that grows in the presence of oxygen.

Anaerobic: A bacterial species that cannot grow in the presence of oxygen. It can be assumed that oxygen is toxic to the cell.

Microaerophilic: A bacterial species that grows poorly (but does grow) in the presence of oxygen. These bacteria grow better in oxygen-poor or oxygen-free environments.

Facultative: A bacterial species that grows well under both aerobic and anaerobic conditions. If this word is used as an adjective (as "facultative aerobe" and "facultative anaerobe"), confusing redundancy may be created. How would you differentiate between a "facultative aerobe" and a "facultative anaerobe"? Choose your words carefully.

Obligate: The term obligate may be used as an adjective to imply an absolute requirement for the environmental condition (as "obligate aerobe" and "obligate anaerobe"). When used in this way, the adjectives *obligate* and *facultative* are, by definition, mutually exclusive.

TECHNIQUES FOR THE CULTURE OF ANAEROBIC BACTERIA

Techniques used for the cultivation of anaerobes depend upon the removal of free oxygen from the environment. The environment includes both the medium and the atmosphere above the medium. Strict obligate anaerobes require that virtually all molecular oxygen be removed from the medium and from the atmosphere above the medium. Remember that oxygen diffuses easily and that any oxygen remaining in the atmosphere will rapidly dissolve in the medium and diffuse throughout. Strict anaerobiosis is difficult to obtain without special media and equipment.

There are many techniques that reliably remove most of the oxygen from the environment. There are a few techniques that reliably remove enough oxygen to allow growth of strict anaerobes. We will examine two of the techniques commonly used in clinical laboratories, as well as one method commonly used to support the growth of microaerophilics.

Thioglycollate Medium

Thioglycollate medium is a nutrient medium that contains thioglycolic acid. This compound is a very strong reducing agent that reacts quickly with any oxygen that may diffuse into the medium. Of course, once the thioglycolic acid has been saturated with oxygen, the medium is no longer useful for the cultivation of anaerobes.

Thioglycollate medium is typically used as a semi-solid, tubed medium and is inoculated with a single stab into the center of the deep. It may, however, be used as a broth, or may be solidified with additional agar and used in plates. The use of a small amount of agar to make it semi-solid has the advantage of reducing the diffusion rate of oxygen.

Some commercially available formulas for thioglycollate medium include an indicator dye that is colored in the presence of oxygen. The dye, typically methylene blue, appears blue when oxidized and is colorless when reduced. The use of such a dye has the obvious advantage of providing a convenient means of monitoring the anaerobiosis of the medium. However, there is evidence that the dye may inhibit the growth of some bacteria, especially certain gram-negative rods, and thioglycollate medium without indicator is used when these bacteria may be present.

When thioglycollate medium is correctly inoculated with a single stab down the center of the deep, the pattern of growth produced is distinct for each type of oxygen-related requirement (see **Figure 9.1**). Obligate aerobic bacteria will grow only in the aerobic zone (colored by dye, near the air/medium surface); obligate anaerobes will grow only in the anaerobic zone (uncolored by dye, in the bottom of the tube); microaerophiles will grow slightly below the surface (primarily in the area of the colored/uncolored interface); and facultative bacteria will grow along the entire length of the stab.

As noted earlier, oxygen dissolves readily in most media, including thioglycollate medium, and will quickly saturate the thioglycolic acid. Oxygen, being a gas, is also easily removed from solution by heating. If the colored portion of the tube extends more than one quarter of the depth of the medium, the medium must be heated to boiling to remove the dissolved oxygen prior to its inoculation. (Why wouldn't you want to boil it following inoculation?) The colored band will disappear upon heating, indicating that the oxygen has been removed. Many laboratories automatically include this step whenever thioglycollate medium is used. It would be essential if you use thioglycollate medium without indicator.

Anaerobic Chambers or Jars

An anaerobic jar is a container that can be rendered anaerobic and which will remain anaerobic as long as it remains correctly sealed. Anaerobiosis is accomplished by using a compound, in this case hydrogen gas, that reacts with the oxygen inside the jar, usually reducing it to water. Once the atmospheric oxygen is depleted, the oxygen in the medium will rapidly diffuse into the atmosphere and, in turn, be reduced. We will use the Gas-Pak anaerobic system as an example. (Gas-Pak is a commercial brand name for a product of Baltimore Biological Laboratories, Inc.) The components of the GasPak system (shown in **Figure 9.2**) are:

1. **Anaerobic jar.** The container consists of the jar itself, a cover, and a clamp that holds the cover tightly on the lip of the jar. Sometimes a sealant, such as a thin layer of petroleum jelly or stopcock grease, may be used to ensure a good, gastight seal between the jar and the cover.

2. **Catalyst.** The catalyst safely increases the rate of reaction between oxygen and the hydrogen gas mixture that is used to reduce it. The Gas-Pak system uses a catalyst that works safely at incubator temperature (in some older systems it was necessary to heat the catalyst electrically). It is usually contained in a small, screened container that is attached to the inner surface of the cover. This catalyst must be replaced occasionally. It is also possible to get gas generator envelopes which contain the catalyst on the outside of the envelope. If these

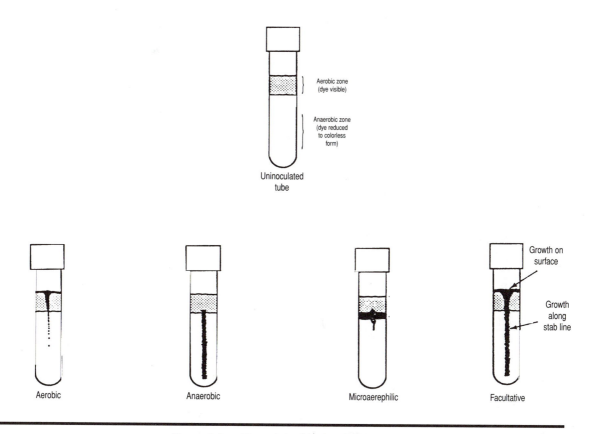

Figure 9.1   Growth in thioglycolate agar.

envelopes are used, additional catalyst in the cover of the jar is not needed.

3. **Gas generator.** The hydrogen-gas mixture that is used to reduce the oxygen is generated by chemicals provided in a foil envelope. It is only necessary to cut a corner off the envelope, add 10.0 ml of water, and place the envelope (with the water inside) into the GasPak jar. The reaction that takes place in the jar produces the gas that will be used to reduce the oxygen. Each Gas-Pak envelope may be used only once.

4. **Indicator.** A disposable indicator is available for use in the Gas-Pak system. It consists of a piece of paper saturated with a methylene blue solution. The methylene blue will turn colorless when the oxygen is removed from the atmosphere inside the jar. Although an indicator is not essential for the operation of the anaerobic system, it should be used to ensure that anaerobiosis was attained. (How else would you be able to determine if the catalyst was

still effective and proper anaerobic conditions were achieved and maintained?)

The Gas-Pak system operates in the following manner:

1. Obtain a Gas-Pak jar and ensure that the surfaces on the jar and cover that will mate when the jar is closed are clean and free of deep scratches and defects. Apply a very light coating of petroleum jelly or stopcock grease to one of the surfaces.

2. Check to ensure that a container of catalyst is attached to the inner surface of the cover, unless it is contained on the gas generator envelope.

3. Loosely pack one or two paper towels into the bottom of the anaerobic jar. Moisture will be produced by the catalyzed reaction; the paper toweling will absorb most of the moisture. This is not necessary if you are using a basket that fits inside the jar to contain the petri plates.

4. Stack the petri plates into the jar, or the basket that comes with the jar, as directed by your instructor. Be careful that you do not try to use too many dishes. When closed, there should be a space of at least one inch between the top petri plate and the catalyst container.

5. Obtain a gas-generating envelope and cut the corner off along the dashed line (found on the upper corner of the envelope). Place the envelope in the jar, either standing it up between the wall of the jar and the stacked petri plates, or in the holder provided on the basket.

6. If you are using an indicator, place the exposed indicator in the jar in such a way that you will be able to observe it through the side of the jar, or in the holder provided on the basket. The paper will initially appear blue (why?). It should turn white when anaerobic conditions have been established.

7. Add 10.0 ml of water to the gas-generator envelope. The tip of the pipette must not be forced into the envelope. You only need to insert the tip far enough to be sure that all the water flows into the envelope.

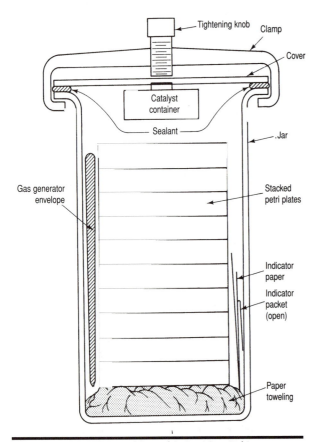

Figure 9.2   Anaerobic jar (Gas-Pak).

8. Quickly place the cover on the jar and clamp it in place. The tightening knob on the clamp should be lightly tightened, just enough to make a good seal between the jar and the cover. Do not overtighten the clamp.

9. Place the closed unit in the incubator and incubate for an appropriate period. Some anaerobic cultures require up to 48 hours of incubation. The Gas-Pak jar must not be opened during the incubation period.

CANDLE JAR

In order to grow microaerophiles in a non-reducing medium, a candle jar is used (see **Figure 9.3**). Inoculated plates or tubes are placed in a jar. A candle is placed in the jar and lit, and the lid of the jar is securely placed on the jar. As the candle continues to burn, the oxygen concentration is diminished and carbon dioxide concentration is increased until there is no longer adequate oxygen to support the combustion of the candle and it is extinguished. The reduced oxygen conditions will be maintained in the jar as long as it remains unopened.

LABORATORY OBJECTIVES

The culture of anaerobic and microaerophilic bacteria requires specialized techniques for the removal of free oxygen. Many pathogenic bacteria are anaerobic or microaerophilic, and their isolation and identification is essential. To understand the nature of anaerobiosis and to appreciate the special techniques used, you should:

- Understand the meaning of the terms *anaerobic, microaerophilic, facultative, and obligate,* as used in this context.

- Be familiar with simple *anaerobic techniques,* including the *thioglycollate medium* and *anaerobic jars.*

- Be familiar with *microaerophilic techniques,* including the *thioglycollate medium* and *candle jars.*

- Understand the significance of *facultative* versus *obligate* organisms in terms of the environmental conditions under which they can grow.

MATERIALS NEEDED FOR THIS LABORATORY

This exercise may be performed individually or in groups as indicated by your instructor.

1. Twenty-four to 48-hour cultures of the following organisms:
   Clostridium sporogenes
   Pseudomonas aeruginosa
   Staphylococcus aureus
   Micrococcus luteus

2. Five screw-capped tubes (deeps) of thioglycollate medium. If the dye is visible for more than one quarter of the depth of the agar deep, the tubes will need to be heated in a boiling water bath to drive off the dissolved oxygen. After you heat the tubes, do not shake them or disturb them while they are cooling.

3. Three plates of nutrient agar

4. Gas-Pak anaerobic jar, with catalyst, indicator, and gas generator envelope

5. 10.0 ml pipette

6. Candle jar and candle

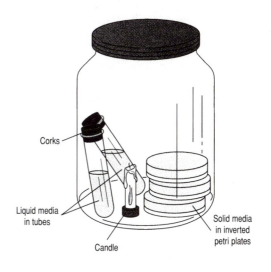

Figure 9.3   Candle jar.

LABORATORY PROCEDURE

1. Obtain one tube of thioglycollate medium for each organism to be tested, plus one tube for use as a control.

2. Label all the tubes and mark the depth of the colored area on the control tube with a marker.

3. Using your sterile inoculating needle, inoculate each tube with a test organism by making a single stab down the center of the deep.

4. Divide three petri plates into quadrants as shown in **Figure 9.4.** Each quadrant will be inoculated with one of the known organisms. Label one plate "aerobic," one plate "anaerobic," and one plate "candle jar."

5. Set up one Gas-Pak anaerobic jar according to the instruction given in this exercise. Place the plates labeled "anaerobic" in the jar. Follow the previous directions for charging the jar. Place the jar in the 37°C incubator. **NOTE:** One jar may be used for more than one set of plates.

6. Place the petri plates labeled "candle jar" in the candle jar. Place the candle on top of the plates and light it. Securely attach the lid. When the candle is extinguished, place the jar in the 37°C incubator.

7. Place the plates labeled "aerobic" directly in the 37°C incubator.

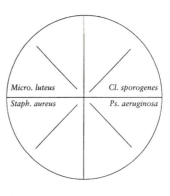

Figure 9.4   Inoculation of Plates

8. Incubate all cultures for 24 to 48 hrs. It is important that the anaerobic and candle jars remain sealed for the entire incubation period. (Why can't you open the jars to check your cultures?)

9. Record the growth pattern observed in the thioglycollate medium. Be sure to note the position of any visible growth relative to the colored area. Measure the change in the depth of the aerobic zone. Record this on the Laboratory Report Form.

10. Examine the plates that were incubated aerobically, anaerobically and microaerophilically. For each organism, compare the amount of growth under each condition, recording growth as (0), (+), (+ +), (+ + +), or (+ + + +). Record your results on the Laboratory Report Form.

LABORATORY REPORT FORM

EXERCISE 9
OXYGEN REQUIREMENTS OF MICROORGANISMS

What was the purpose of this exercise?

INDICATE YOUR RESULTS FOR THE FOLLOWING:

1. Thioglycollate Medium: For each organism, indicate the location of growth and the depth of the aerobic zone.

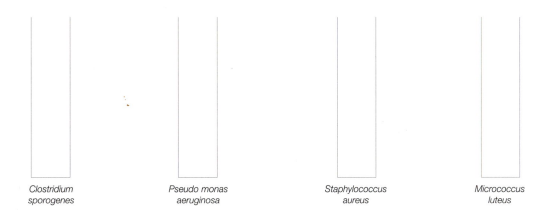

2. Petri Plates: Indicate the comparative growth for each organism in the following table. Under *Oxygen Requirements,* indicate whether the organism appears to be an obligate aerobe, obligate anaerobe, facultative, or microaerophilic.

ORGANISM	AEROBIC (0, +, ++, +++, ++++)	ANAEROBIC JAR (0, +, ++, +++,	CANDLE JAR (0, +, ++, +++, ++++)	OXYGEN REQUIREMENTS
Clostridium sporogenes				
Pseudomonas aeruginosa				
Staphylococcus aureus				
Micrococcus luteus				

3. Do your results for the thioglycollate medium and the plates agree? If not, what explanation might account for the variation?

QUESTIONS:

1. List two pathogenic species of *Clostridium* and the disease that each causes.

a.

b.

2. List two other pathogenic bacteria, not members of the genus *Clostridium,* that are anaerobic.

a.

b.

3. List two energy-producing pathways that do not require the utilization of molecular oxygen.

4. Humans are considered aerobic organisms. If so, why do we need to worry about the culturing of anaerobic or microaerophilic organisms?

PHYSICAL GROWTH REQUIREMENTS

Microorganisms are said to be *ubiquitous,* present almost everywhere. This would imply their ability to grow under environmental conditions exhibiting wide variations in temperature, osmotic pressure and pH. To not only survive, but to often flourish under these varied conditions, many organisms have developed widely varied adaptive mechanisms. An organism may grow optimally at 25°C, but still be able to survive at 4°C. Other organisms are able to grow in hot springs, a pickle brine, or vinegar. Even so, at some point an organism's adaptive mechanisms for temperature regulation, osmotic pressure, or pH can be overcome and vital enzymes or other cell constituents will no longer be able to function properly.

In this exercise we will be examining not only the optimal temperature, osmotic pressure and pH conditions for several microbial species, but also the varied conditions under which they can survive. Why would it be important to know this information about given organisms?

BACKGROUND

TEMPERATURE

The rate at which chemical reactions take place in a cell is determined by enzyme activity. That temperature at which a cell's enzymes function optimally is referred to as the ***optimal growth temperature.*** As the temperature of the cell is decreased from its optimum, the rate of enzymatic activity will slow at a rate of approximately 50% for every 10°C drop in temperature. Increased temperatures can result in the

irreversible denaturing of the enzyme and therefore the cessation of all activity. **Figure 10.1** illustrates the relationship between temperature and microbial growth. The ***minimum growth temperature*** is the lowest temperature at which the species will grow; conversely, the ***maximum growth temperature*** is the highest temperature at which it can grow. Of course, the *optimum growth temperature* is that at which it grows best.

Bacteria are divided into three major groups based on the temperature at which they grow optimally. Those organisms whose growth range falls between -5°C and 20°C are referred to as ***psychrophiles.*** Their optimum temperature is around 15°C. These organisms are commonly found in the oceans or in refrigerators where they may be responsible for food spoilage.

Mesophiles are those organisms with optimum growth temperatures between 25°C and 40°C, with many of them growing optimally at 37°C, or human body temperature. Those organisms which comprise the normal flora of humans, as well as human pathogens, are mesophilic. Also included in this group are many of the organisms commonly found in soil. A special group of the mesophiles are the ***psychrotrophs,*** or those organisms which can also grow at temperatures as low as 0°C. The majority of the organisms we have used or will use in the laboratory are mesophiles.

The final group of organisms based on optimum temperatures are the ***thermophiles.*** These organisms will have a growth range of from 45°C to 65°C although some are able to grow in temperatures

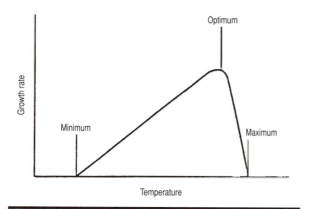

Figure 10.1   Growth curve.

greater than 90°C. These organisms are often isolated from hot springs, deserts, or compost piles.

It must be noted that temperature classification (see **Figure 10.2**) is only a measure of the optimal temperature for an organism's growth and is not reflective of its ability to survive. Endospore-forming organisms, for example, are able to survive in a dormant state over very wide temperature ranges.

The effect of temperature on microbial enzymes can be seen in ways other than growth rates. *Serratia marcescens*, for example, produces a temperature-dependent red-orange pigment, prodigiosin, only at specific temperatures.

In addition to determining the optimum growth temperatures for each organism, it is important to know its **thermal death time** and **thermal death point.** The former is the time it takes to kill an organism at a given temperature, while the latter is the temperature necessary to kill an organism in 10 minutes. This information is important in the determination of the temperature and time conditions necessary for killing a given organism. You will determine the thermal death time for an organisms in Exercise 21.

OSMOTIC PRESSURE

Osmotic pressure is a measure of the force water exerts on a semipermeable cell membrane due to variations in the concentration of dissolved solutes in the medium or cytoplasm on either side of the cell membrane. If the solute concentration is less outside the cell than it is within the cell, water will tend to flow into the cell through the process known as *osmosis.* This movement of water from the point of lower solute concentration to that of higher solute

concentration can facilitate the movement of nutrients into the cell as well as maintain the internal cell pressure.

Those environments which exhibit lower solute concentrations than are found within the cell are referred to as being *hypotonic.* Because of the presence of its rigid cell wall, bacterial organisms are not as sensitive to hypotonic solutions as are cells, such as red blood cells, which rapidly undergo hemolysis or rupture when placed in hypotonic solutions.

If the solute concentration is lower inside the cell than it is outside the cell, water will tend to flow out of the cell, resulting in dehydration of the cell as well as possible denaturing of the proteins within the cell due to the high salt concentration that becomes present. These solutions are *hypertonic* to the cell and often cause cell death. Often hypertonic solutions are used in food preservation. What would be some examples?

Most bacteria grow best in solutions containing solute concentrations close to that of cytoplasm (*isotonic* solutions). Other organisms, however, have had to adapt to existing in environments that are hypertonic. Some *Staphylococcus* strains will survive in salt concentrations up to 10%. This is important as they inhabit the skin of humans which, due to the evaporation of perspiration, can become relatively salty. (How does it feel when you get perspiration in your eyes?) Other organisms have become so well adapted to elevated salt concentrations that they cannot survive at lower salt concentrations. These organisms can be isolated from such environments as the ocean (salt concentration approximately 3%), salt lakes (15 – 30% salt concentration), or pickle brines (greater than 15% salt concentration). Organisms requiring a salt concentration of 0.2M NaCI or higher are *halophiles* and can only be grown in higher salt concentrations.

Many of the fungi are *saccharophiles.* What dissolved substance can they withstand in far higher concentrations than can other organisms? How is jelly preserved? Can any microorganism grow on its surface?

pH

pH is a measurement of hydrogen ion concentration in a solution. Solutions with increased hydrogen ions are referred to as being acidic; those with decreased levels are basic. Most bacteria are *neutraphiles,* preferring a pH range of 6.5 – 7.5. As is

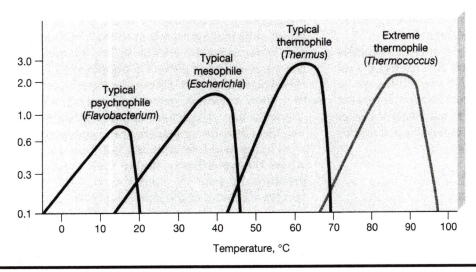

Figure 10.2   Temperature growth ranges for psychrophilic, mesophilic and thermophilic organisms.

the case with both temperature and osmotic pressure, we can see the adaptability of microorganisms as we look at the environments in which they survive. For example, skin organisms must be able to withstand a pH of around 5 and those organisms of the gastrointestinal tract can be exposed to pH levels of 3 or below. Those organisms which can tolerate lower pH values, but do not actively grow at those levels, are referred to as being **acidoduric.** Those that not only tolerate, but readily grow under these conditions are **acidophiles.** The majority of the acidophilic microorganisms are fungi (fungi have an optimal pH range of 4 – 6). How could you use this information in developing a media to selectively grow the fungi?

The pH of growth medium is altered by the metabolic activity of organisms growing in it. Sugar fermentation can result in the production of lactic acid. Protein degradation produces amines and ammonia. Either or these reactions could result in the inhibition of bacterial growth due to increased acidity (lactic acid) or alkalinity (ammonia) in the medium. Buffers may be added to media to prevent the resultant change in overall pH of the medium .

LABORATORY OBJECTIVES

Each microorganism must be adapted to live in its natural habitat, in spite of what often appears to be adverse conditions of temperature, osmotic pressure or pH. To better understand the adaptations that they have evolved we must:

- Compare and contrast the natural environments of *psychrophiles, mesophiles, and thermophiles.*

- Determine the effect of temperature on bacterial growth.

- Define *osmotic pressure* and explain how it affects a cell.

- Compare the effects on microbial cells of *isotonic, hypertonic,* and *hypotonic* solutions.

- Explain how microbial growth is related to pH.

MATERIALS NEEDED FOR THIS LABORATORY

A. TEMPERATURE

1. TSA Plates—8

2. 24-hour cultures of
    Pseudomonas fluorescens
    Escherichia coli
    Staphylococcus aureus
    Bacillus sterothermophilus
    Serratia marcescens

3. Incubators:
    Refrigerator (4°C)
    37°C
    55°C

B. OSMOTIC PRESSURE

1. TSA Plates—1 each

TSA (0% NaCl + 0% Sucrose)
TSA + 5% NaCl
TSA + 10% NaCl
TSA + 15% NaCl
TSA + 10% Sucrose
TSA + 25% Sucrose
TSA + 50% Sucrose

2. 24-hour cultures of
   Escherichia coli
   Staphylococcus aureus
   Saccharomyces cerevisiae

3. Spore suspension of *Penicillium*

C. pH

1. Nutrient broth with pH adjusted to 3.0, 5.0, 7.0, and 9.0 (4 each)

2. 24-hour cultures of
   Escherichia coli
   Staphylococcus aureus
   Saccharomyces cerevisiae
   Lactobacillus delbruckii subspecies *bulgaricus*

LABORATORY PROCEDURE

A. TEMPERATURE

1. Divide one TSA plate for each incubation temperature into 4 segments. Label 1 section on each plate with each of the following organisms:
   Pseudomonas fluorescens
   Escherichia coli
   Staphylococcus aureus
   Bacillus sterothermophilus
   Also include your identifying information and the incubation temperature.

2. Using a sterile loop, inoculate the center of each section with a short straight line of the appropriate organism.

3. Incubate each plate at the designated temperature for 24 – 48 hours.

4. Label the remaining plates for inoculation with *Serratia marcescens*. Be sure to include the incubation temperature on each.

5. Using a sterile loop, prepare a streak plate with *Serratia marcescens* on each.

6. Incubate each plate at the designated temperature for 24 – 48 hours.

7. Compare the amount of growth of each organism at the varied temperatures. Be sure to compare the amount of growth for a given organism at each temperature. Record the growth as (0), (+), (+ +), (+ + +), or (+ + + +).

8. Compare the pigment production of the *Serratia marcescens*, noting both the color and its intensity at the varied temperatures.

9. Record your results on the Laboratory Report Form, Part A.

B. OSMOTIC PRESSURE

1. Divide each plate into 4 segments. Label each quadrant with one of the organisms.

2. Using your sterile loop, inoculate *Escherichia coli* with a short straight line onto each plate. Repeat with *Staphylococcus aureus* and *Saccharomyces cerevisiae*. Inoculate *Penicillium* in a small circular area (no larger than dime size) in the center of the fourth segment on each plate.

3. Incubate the plates for 24 – 48 hours. Compare the relative amounts of growth of each organism under the varied osmotic conditions. Record the growth as (0), (+), (+ +), (+ + +), or (+ + + +) on the Laboratory Report Form.

4. Continue to incubate the plates at room temperature an additional 3 – 5 days. Record the relative amounts of mold growth on the Laboratory Report Form, Part B.

C. pH

1. Inoculate each organism into one tube of TS broth at each of the pHs.

2. Incubate the tubes for 24 – 48 hours.

3. Resuspend the organisms in each broth. Determine the relative amounts of growth, comparing each organism for each of the pHs. Record the results as (0), (+), (+ +), (+ + +), or (+ + + +) on the Laboratory Report Form, Part C.

LABORATORY REPORT FORM

EXERCISE 10
PHYSICAL GROWTH REQUIREMENTS

What was the purpose of this exercise?

A. TEMPERATURE

1. What is the ambient temperature of the laboratory?

2. Complete the following table:

Effect of Temperature on Growth

ORGANISM	4°C	AMBIENT TEMPERATURE	37°C	55°C
Pseudomonas fluorescens				
Escherichia coli				
Staphylococcus aureus				
Bacillus sterothermophilus				

3. Compare the appearance of *Serratia marcescens* at the varied temperatures.

	4°C	AMBIENT TEMPERATURE	37°C	55°C
AMOUNT OF GROWTH				
APPEARANCE				

4. Attempt to classify each of the following organisms as psychrophilic, mesophilic, or thermophilic.

   Pseudomonas aeruginosa _____

   Escherichia coli _____

   Staphylococcus aureus _____

   Bacilus sterothermophilus _____

B. OSMOTIC PRESSURE

1. Record the following tables with your results:

Effect of NaCl Concentration on Microbial Growth

ORGANISM	0% NaCl	5% NaCl	10% NaCl	15% NaCl
Escherichia coli				
Staphylococcus aureus				
Saccharomyces cerevisiae				
Penicillum				

Effect of Sucrose Concentration on Microbial Growth

ORGANISM	0% Sucrose	10% Sucrose	25% Sucrose	50% Sucrose
Escherichia coli				
Staphylococcus aureus				
Saccharomyces cerevisiae				
Penicillum				

C. pH

1. For each of the following organisms, give its natural habitat. What pH would you expect to be its optimum?

ORGANISM	NATURAL HABITAT	PREDICTED OPTIMUM pH
Escherichia coli		
Staphylococcus aureus		
Saccharomyces cerevisiae		
Lactobacillus delbruckii subspecies *bulgaricus*		

2. For each of the following organisms, report its relative growth at the various pHs.

ORGANISM	pH 3.0	pH 5.0	pH 7.0	pH 9.0
Escherichia coli				
Staphylococcus aureus				
Saccharomyces cerevisiae				
Lactobacillus delbruckii subspecies *bulgaricus*				

QUESTIONS:

1. An organism is inoculated onto TSA plates and incubated at 4°C, 23°C, 37°C and 55°C for 48 hours. The following results are noted

4°C	23°C	37°C	55°C
+ +	+ +	+ ++ +	0

   How would you classify this organism based on temperature of growth?

2. Based on your experimental results, what is the medical significance of endospore-forming organisms?

3. Why was it necessary to inoculate a plate containing 0% NaCl and 0% sucrose?

4. What organisms are most likely to be involved in the spoilage of jelly?

5. Compare the osmotic tolerance of bacteria and fungi.

6. You are attempting to grow large quantities of an organism, but it produces large quantities of lactic acid, lowering the pH to the point that the organisms cannot survive. How could you prevent this?

7. *Heliobacter pylori* grow in the stomach. They also produce large quantities of urease. Why?

SELECTIVE AND DIFFERENTIAL MEDIA

We saw in Exercises 9 and 10 that different microorganisms have varied physical growth requirements which must be met if we are to successfully culture them. In addition to their physical growth requirements, each organism also has specific nutritional requirements. For some species, they will only need to have a surface to grow on, limited inorganic salts, and air. Others are dependent on the provision of complex organic molecules to meet their nutritional needs.

Media can also be utilized to provide us with additional information about a given organism. By adding pH indicators, alternative sources of carbon or nitrogen, or inhibitory chemical agents we can determine the varied nutrients an organism is capable of utilizing, the by-products of metabolism produced, or the organism's ability to withstand adverse chemical agents.

BACKGROUND

Bacteriological media can serve many purposes in the microbiology laboratory—from enabling the collection of specimens, transport of specimen to the laboratory, detection, primary isolation and identification of microorganisms, and the determination of antimicrobic susceptibility. In this exercise we will concentrate on those media which aid in the primary isolation and identification of microorganisms.

We will only be exploring the utilization of a few of the hundreds of types of media available to the microbiologist. The choice of media will depend upon the source of the organism or specific organisms being screened for. Complete information about the composition and uses of varied medias can be found in *Difco Manual* (Difco Laboratories) and the *BBL Manual* (Becton-Dickinson).

TRANSPORT MEDIA

One of the most critical factors in clinical microbiology is the adequate preservation of the organism from the time it is collected from the patient to the the time of culturing in the laboratory. Transport media are non-nutrient, semi-solid medias which inhibit self-destructive enzymatic reactions within the cell. It also is formulated to prevent the lethal effects of oxidation. Examples of transport media include Amies Transport Media and Stuart Transport Media. Transport media will be discussed further in Exercise 29.

ISOLATION MEDIA

The purpose of isolation media is to determine the presence of microorganisms in the culture material. Often this requires the elimination of normal flora organisms so the presence of pathogens can be determined. When it comes to choosing a selective media, several criteria must be considered. These include the source of the specimen, what transport media was used, the Gram reaction of the culture, the presence of commensal organisms, and the nature of the suspected agent.

General Purpose Media

Nonselective media used for primary isolation of microorganisms are the general purpose media. These media will support the growth of a wide variety of normal body flora, pathogens and soil microbiota. To encourage the growth of organisms, enrichment factors, such as blood, hemoglobin and growth factors, may be added. Examples of general purpose media include Nutrient Agar, Tryptic Soy Agar, and Blood Agar for bacterial growth and Sabouraud Dextrose Agar for fungi.

Selective Media

Often, when attempting to isolate organisms, we are not interested in those organisms which compose the normal flora of the environment, but instead are looking for the presence of other species which are not routinely present. This could include screening for the presence of *Escherichia coli* in a body of water (indicative of water contamination with fecal material) or for the presence of mycobacteria in a sputum sample. To facilitate the detection of these organisms amongst the normal flora, selective media are used.

Selective media is formulated to suppress or inhibit the growth of one group of microorganisms while allowing the growth of others present in a mixed flora. This is usually achieved by the incorporation of bacteriostatic agents or the alteration of the physical or chemical environment of the media. Examples would include the inclusion of crystal violet which, in concentrations of 1:100,000, is inhibitory to gram-positive organisms but not to gram-negative organisms. Other dyes, antibiotics, sodium chloride, sodium azide, phenylethanol, and bile salts are just a few of the commonly used selective agents. Examples of selective media include Columbia CNA Agar, Phenylethanol Medium, and Lowenstein-Jensen Agar.

Enrichment Media

At times, the presence of large numbers of normal flora can overgrow and obscure the presence of pathogens. Enrichment media is designed to suppress the competing normal flora, enabling the growth and further isolation of the pathogen. Examples include GN Broth for the selective cultivation of *Salmonella* and *Shigella* from stool specimens and Selenite Cystine Broth for the enrichment of *Salmonella* from food and dairy products.

Specialized Isolation Media

Specialized isolation media are formulated to satisfy the needs of a specific microorganism, thereby allowing its more rapid isolation and identification. An example would be Mannitol Salt Agar which is selective for the pathogenic staphylococci.

IDENTIFICATION MEDIA

Once an organism has been isolated, it is important that its identity be speedily and accurately determined. This enables the appropriate treatment of infections or elimination of further entry by undesired microorganisms into the environment. Identification media often utilize knowledge of specific enzyme systems to enable more rapid identification.

Differential Media

A differential media incorporates certain reagents or chemicals into the media which result in recognizable reactions following incubation. For example, if a fermentable sugar, such as glucose, is present in the media as well as a pH indicator it will provide, following incubation, a quick visual distinction between those organisms which are fermenters and those which are non-fermenters.

It is possible for a single medium to be both a general purpose medium and a differential medium. We have already discussed Blood Agar as a general purpose medium, capable of supporting the growth of a large number of microorganisms. Certain organisms, however, cause specific visible changes in the media due to their varied production of hemolysins (enzymes which cause the rupture of red blood cells). If the organisms are able to lyse the blood cells present, it results in a complete clearing of the agar around the colony—a process referred to as beta-hemolysis. If this is present on a Blood Agar plate inoculated with a throat culture it is highly significant for the presence of the beta-hemolytic streptococci or "strept throat."

Other media have been formulated to be both selective and differential—providing chemical agents, such as dyes, that will inhibit the growth of many organisms and of also providing other chemicals, such as fermentable carbohydrates, that will enable the differentiation between those that are growing. An example of this is Eosin Methylene Blue (EMB) Agar. This medium is commonly used for the detection of fecal organisms in water supplies. A free-standing body of water will contain many different organisms, most of which are derived from soil. The dyes eosin and methylene blue are inhibitory to the growth of those Gram-positive cells that are present in the water. Many of the Gram-negative organisms that are present, such as *Enterobacter aerogenes,* are found in soil run-off and need to be distinguished from those of possible fecal origin, such as *Escherichia coli.* By incorporating lactose ("milk sugar"), which is not fermented by most soil organisms, it is possible to presumptively determine the identity of *E. coli* based on the production of acid (seen as a green metallic sheen).

OBJECTIVES

- Define all-purpose, selective, and differential media.

- Distinguish between bacteria by utilizing all-purpose, selective, and differential media.

- Demonstrate and judge the advantages of a medium that is both selective and differential.

- Understand the relationship between the growth of an organism and the composition of the medium.

- Learn the advantages in the use of selective, differential, and enriched media in the laboratory cultivation of microorganisms.

MATERIALS NEEDED FOR THIS LABORATORY

1. Broth cultures of
   Escherichia coli
   Enterobacter aerogenes
   Staphylococcus aureus
   Streptococcus faecalis

2. Media, 1 per group
   Tryptic Soy Agar plate
   Phenylethanol Agar plate
   Mannitol Salt Agar plate
   Levine EMB plate
   Blood Agar plate
   Columbia CNA plate

PROCEDURES

1. Label each plate with the name of the media and your identifying information. Divide each plate into quadrants. Label each section with the name of one of the test organisms. (See **Figure 11.1**).

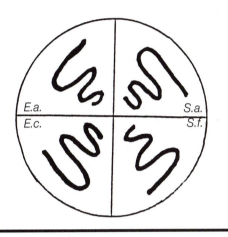

Figure 11.1   Inoculation of agar plates.

2. Inoculate each section with the appropriate organism.

3. Incubate all plates at 37°C.

4. Following incubation, record the results for each of the plates on the Laboratory Report Form:
   a. Tryptic Soy Agar: observe the plates for the presence of growth. What would it mean if there were no growth on one of the sections?
   b. Phenylethanol Agar: Observe the plate for the presence of growth.
   c. Mannitol Salt Agar: Observe the plate for the presence of growth. For those organisms that grew, record the color of the surrounding medium (pink or yellow).
   d. Levine EMB Agar: Observe the plate for the presence of growth. For those organisms that grew, record the color of the colonies (pink, metallic green, or colorless).
   e. Blood Agar: Observe the plate for the presence of growth. For those organisms which grew, record the hemolysis present

as α-hemolysis (zone of greening around the growth), β-hemolysis (zone of clearing around the growth), or γ-hemolysis (absence of hemolysis).

f. Columbia CNA Agar: Observe the plate for the presence of growth. For those organisms that grew, record the type of hemolysis observed.

Name ———————————

Section ———————————

LABORATORY REPORT FORM

EXERCISE 11
SELECTIVE AND DIFFERENTIAL MEDIA

What was the purpose of this exercise?

1. Record your results on the following chart:

MEDIUM	*Escherichia coli*	*Enterobacter aerogenes*	*Staphylococcus aureus*	*Streptococcus faecalis*
Tryptic Soy Agar Growth				
Phenylethanol Agar Growth				
Mannitol Salt Agar Growth				
Appearance				
Levine EMB Agar Growth				
Appearance				
Blood Agar Hemolysis				
Columbia CNA Agar Growth				
Appearance				

2. Based on your results, which media would you consider to be:

    a. General purpose:

    b. Selective:

    c. Differential:

    d. Selective and Differential:

QUESTIONS:

1. Describe the basis for selectivity of EMB agar, MacConkey agar, and Mannitol Salt agar.

2. You are working in a clinical laboratory. When might you choose to use the following media? (You may find it helpful to read about each of the media in *Difco Manual*)

    a. Desoxycholate Medium:

    b. Phenylethyl Alcohol Medium:

    c. Blood Agar:

3. A urine specimen is sent to the clinical laboratory indicating that it is from a patient with a urinary tract infection. The microbiologist plates it out on MacConkey Agar. Why?

4. When you go to the doctor's office with pharygitis (a sore throat), they will often take a swab of your throat and plate it on Blood Agar. Why?

PHYSIOLOGICAL CHARACTERISTICS OF BACTERIA:
CARBOHYDRATE METABOLISM

The next three exercises cover the physiological characteristics of bacteria.

There are at least two reasons for studying the biochemical characteristics of bacteria. First, the characteristics can be used to demonstrate the exceptional metabolic diversity of prokaryotic organisms. The range of metabolic capabilities that bacteria exhibit is very large, explaining in part why these organisms can be found in virtually every environmental habitat. Not only are bacteria capable of obtaining energy by a variety of pathways, some of which are unique to bacteria, but they are capable of utilizing a very large number of different metabolites.

The second reason is that the biochemical characteristics of bacteria represent additional phenotypic characteristics that can be easily examined. The fact that individual species characteristics are genetically determined makes it possible to use them as phenotypic markers. These biochemical characteristics make it possible to identify unknowns by matching the phenotype of the unknown to that of a known reference organism. A reference organism is one that is considered to be typical of a given genus and species—so much so that it can be used as a reference against which unknown isolates can be compared.

BACKGROUND

The reactions of carbohydrates that are employed for identification are usually degradative catabolic reactions used by the bacteria as part of their energy-producing metabolism. The synthesis of structural and storage carbohydrates is also important, but these reactions are not often used in identification protocols. The reactions you will study fall into two major categories: those which determine *if* certain carbohydrates are used at all, and those which determine *how* they are used. For example, we will test several bacterial species to determine which sugars they are able to use (i.e., glucose, lactose, etc.) and whether or not the sugars can be fermented with the production of acid and gas, or only acid.

A NOTE ABOUT pH INDICATORS

pH indicators will change color as the pH of the solution they are in changes. The actual color change takes place over a range, the center of which is referred to as the pK. The indicators you will use in this exercise are:

Indicator	pK	Range and Color
Methyl red	5.2	4.4 – 6.0 Red-Yellow
Phenol red	7.9	6.8 – 8.4 Yellow-Red
Brom thymol blue	7.0	6.0 – 7.6 Yellow-Blue

These data indicate that phenol red will be yellow below pH 6.8, that it will begin to turn red at 6.8, and that it will be completely changed by pH 8.4.

101

Between 6.8 and 8.4 it will be various shades of orange, becoming increasingly more red as the pH increases and increasingly more yellow as the pH decreases. Brom thymol blue will change from yellow to green to blue as the pH rises from 6.0 to 7.6.

EXPERIMENTAL METHODS

In order to determine an organism's biochemical characteristics, you must use a medium that induces or enhances that characteristic and you must use some chemical test to measure the activity. The latter often involves the use of pH indicators in the medium to detect the production of either acids or bases, but may also depend upon added chemicals that react with the products to give a colored compound.

Fermentation of Sugars

Sugar-fermentation broths contain the sugar being tested and a pH indicator to detect the production of acid. In this exercise, you will use phenol red sugar broths. If the sugar is fermented, the medium will turn from red to yellow because of the production of acid. The glucose sugar tube will also have a smaller tube inside for the entrapment of gas bubbles, should any gas be produced. This is referred to as a Durham tube. If the bacteria do not utilize the sugar, they will often use the amino acids contained in the medium. When amino acids are used, ammonia is produced as a by-product, causing the pH of the medium to rise (the color will turn deeper red). If the sugar is used oxidatively (respiration), there will be little, if any, color change. Sugar fermentation results are reported as *acid* (A), *acid + gas* (A + G), *alkaline* (Alk), or *no change* (NC).

Hydrolysis of Starch

When iodine and starch react, a deep purple color results. If a starch agar plate is flooded with an iodine solution, any starch present will react with the iodine, causing a deep purple color to form. If the starch has been hydrolyzed around the colonies, the medium will remain unstained. The enzyme secreted by bacteria that hydrolyze starch is called amylase; the test is occasionally referred to as the *amylase test*.

Methyl Red-Voges-Proskauer (MR-VP) Test

Methyl Red (MR) Test

Methyl red is a pH indicator that changes color at a lower pH (about 4.5) than phenol red. When a few drops are added to a broth containing a sugar, a color change will result if the pH falls below about 4.5. You should remember that this indicator turns red below a pH of 4.5, and yellow in a pH above that point. Unlike the phenol red used in the fermentation broths, a red color is positive for the methyl-red test, indicating a pH on the acid side of this indicator. This test is used in most identification schemes for the gram-negative rods.

Voges-Proskauer (V-P) Test

This is a specific test for metabolic intermediates produced when sugars are fermented to produce butylene glycol in addition to acids. Butylene glycol fermentation is one of the fermentation patterns exhibited by some gram-negative rods. it is an alternate fermentation pattern to those that result in the production of acids only. A positive V-P test is almost never observed with a positive M-R test because the M-R test is based on the production of sufficient acid to lower the pH to below 4.5. If some of the sugar being fermented is used to make butylene glycol, then it cannot be used to produce acids. In this test you add Barritt's reagents (A and B) to the medium and observe a color change after allowing sufficient time (about ten minutes) for the reaction to occur.

Citrate Utilization

The medium (Simmon's citrate agar) used to determine if a given isolate is able to utilize citrate has citrate as the only source of oxidizable carbohydrate. It also has a pH indicator that changes color if the citrate is used. The indicator (Brom thymol blue) turns from yellow to blue as the citrate is utilized. The reasons for the pH change, which is from neutral (a green color-- why?) to alkaline, are complicated and related to the reactions of the alkaline earth metals in an aqueous medium. For the purposes of this exercise, it is necessary only to note that as the citrate is used, these reactions will occur and the indicator will change from green to blue.

OBJECTIVES

The objectives of this exercise are related to the reasons given for the study of bacterial biochemical characteristics. You should:

- Understand the difference between fermentation and respiration and why a pH change is indicative of fermentation.

- Obtain an appreciation for the metabolic capabilities of bacteria and how these capabilities can be used to identify bacteria.

- Become introduced to the characterization of bacteria on the basis of their phenotypic characteristics.

MATERIALS NEEDED FOR THIS LABORATORY

This exercise may be performed by groups of students as directed by your laboratory instructor.

A. FERMENTATION OF SUGARS

1. Twenty-four hour cultures of the organisms to be tested, including:
Staphylococcus epidermidis
Escherichia coli
Enterobacter aerogenes
Proteus vulgaris
Bacillus subtilis

2. One tube of each per organism:
Phenol red glucose broth (in a Durham tube)
Phenol red sucrose broth
Phenol red lactose broth
Phenol red mannitol broth

B. HYDROLYSIS OF STARCH

1. Twenty-four hour cultures of the organisms to be tested, including:
Staphylococcus epidermidis
Escherichia coli
Enterobacter aerogenes
Proteus vulgaris
Bacillus subtilis

2. One petri plate of Starch agar per organism.

3. Iodine solution

C. METHYL RED - VOGES-PROSKAUER TEST

1. Twenty-four hour cultures of the organisms to be tested, including:
Staphylococcus epidermidis
Escherichia coli
Enterobacter aerogenes
Proteus vulgaris
Bacillus subtilis

2. One tube of MR-VP broth per organism

3. Sterile test tube

4. Sterile pipet and pipetter

5. Reagents:
Methyl red indicator
Barritt's Reagent A
Barritt's Reagent B

D. CITRATE UTILIZATION

1. Twenty-four hour cultures of the organisms to be tested, including:
Staphylococcus epidermidis
Escherichia coli
Enterobacter aerogenes
Proteus vulgaris
Bacillus subtilis

2. One tube Simmon's citrate agar per organism

LABORATORY PROCEDURE

A. FERMENTATION OF SUGARS

1. Obtain and label the following tubes of media for each organism to be tested: Phenol red glucose broth (in a Durham tube), Phenol red lactose broth, Phenol red sucrose broth, Phenol red mannitol broth

2. Inoculate one of each of the sugar fermentation tubes with each organism. Aseptically transfer a loopful of culture to the tube.

3. Incubate all tubes at 37°C for 24 hours.

4. Following incubation, observe each tube for growth, the production of acid and, in the glucose broth, the production of gas. Record your results on the Laboratory Report Form.

B. HYDROLYSIS OF STARCH

1. Obtain and label one Starch agar plate for each organism to be tested.

2. Using aseptic technique, inoculate each plate with a single streak in the center of the plate.

3. Incubate all plates at 37°C for 24 hours.

4. Following incubation, flood the plates with iodine solution. Allow a few minutes for the color to develop. A clear (not stained purple) area around the colonies indicates that starch has been hydrolyzed. If the purple color extends up to the colony, no hydrolysis occurred. You will find that, if you leave the iodine-flooded plates out in light the purple coloration

will fade. Be sure to record the results shortly after the addition of the iodine.

5. Record your results on the Laboratory Report Form.

C. METHYL RED - VOGES PROSKAUER TEST

1. Obtain and label one tube of MR-VP broth for each organism.

2. Inoculate each tube, being sure to use aseptic technique.

3. Incubate the inoculated tubes at 37°C for 24 hours.

4. Following incubation, transfer 1.0 ml of the MR-VP broth to a sterile test tube.
   a. To the larger volume, add 0.5 ml (about 10 drops) of methyl-red indicator. If the indicator remains red, the test is positive, if it turns yellow, the test is negative.
   b. To the tube containing 1.0 ml of MR-VP broth, add 0.6 ml (about 12 drops) of Barritt's reagent A and 0.2 ml (about 4 drops) of Barritt's reagent B. Mix well and wait ten minutes for the color to develop. A red color indicates a positive test, and a yellow color indicates a negative test.

5. Record the results on the Laboratory Report Form.

D. CITRATE UTILIZATION

1. Obtain and label one tube of Simmon's citrate agar for each organism being tested.

2. Using aseptic technique, inoculate the surface of the slant with your inoculating loop.

3. Incubate all tubes at 37°C for 24 hours.

4. Observe the Simmon's citrate slants for color change. The presence of a blue color on the surface of the slant is a positive test.

5. Record your results on the Laboratory Report Form.

Name _____

Section _____

LABORATORY REPORT FORM

EXERCISE 12
PHYSIOLOGICAL CHARACTERISTICS OF BACTERIA

What was the purpose of this exercise?

Record your results for the following tests:

ORGANISM	GLUCOSE			LACTOSE		SUCROSE		MANNITOL	
	Growth	Acid	Gas	Growth	Acid	Growth	Acid	Growth	Acid
Staphylococcus epidermidis									
Escherichia coli									
Enterobacter aerogenes									
Proteus vulgaris									
Bacillus subtilis									

ORGANISM	STARCH		MR		V-P		CITRATE	
	Color	Reaction (+ or -)	Color	Reaction (+ or -)	Color	Reaction (+ or -)	Color	Reaction (+ or -)
Staphylococcus epidermidis								
Escherichia coli								
Enterobacter aerogenes								
Proteus vulgaris								
Bacillus subtilis								

QUESTIONS:

1. List two reasons for studying biochemical characteristics of bacteria.

    a.

    b.

2. A student observes a tube of phenol red sucrose broth after a 24 hour incubation and notes that the tube is yellow. The tube is left in the incubator. When observed 3 days later, it appears dark red. What happened?

3. How is it possible for an organism to grow on starch agar if it does not hydrolyze starch?

4. Why is it unlikely that an organism would be positive for both the Methyl Red and the Voges-Proskauer test?

PHYSIOLOGICAL CHARACTERISTICS OF BACTERIA, PART II:

REACTIONS OF NITROGEN METABOLISM

In Exercise 12, we started to look at the various ways that bacterial organisms could utilize available nutrients for growth. In this exercise, we will look at how varied organisms can utilize nitrogen-containing compounds.

BACKGROUND

Bacteria normally metabolize carbohydrates as a source of energy, preferring to utilize proteins and amino acids for synthesis and growth. When, however, there is no oxidizable carbohydrate in the medium, or if the bacteria are unable to use the one that is there, the organism must utilize any available amino acids and proteins, oxidizing the carbon chains and often excreting the unused, residual parts of the molecules, such as ammonia and hydrogen sulfide. The following medias restrict the availability of carbohydrates for bacterial growth. This forces the organisms to utilize, when possible, the available proteins and amino acids. We can then test for the by-products of protein decomposition or look for other evidence of proteolytic activity (the hydrolyzing or breakdown of protein).

HYDROLYSIS OF GELATIN

Gelatin is a protein colloid, maintaining its gel state at temperatures below about 25°C. The gel colloidal state is also dependent on the structural integrity of the proteins found in gelatin, and any hydrolysis of the peptides will result in destruction of the colloid. Organisms which produce the enzyme gelatinase can hydrolyze gelatin, with the degree of gelatin hydroly-

sis serving as an indicator of the activity of the enzyme. The resulting products are soluble peptides and amino acids which can serve as a source of carbon and energy for the organism. To test for gelatin liquefaction, it is only necessary to inoculate a tube of nutrient gelatin, incubate for 24 hours, and then determine if the gelatin will solidify when cooled. Failure to solidify (when an uninoculated control has solidified) is a positive test for gelatin hydrolysis.

INDOLE PRODUCTION

Indole is a compound produced when the amino acid tryptophan is hydrolyzed to pyruvic acid (which is used for energy metabolism) and indole (which is excreted) through the action of tryptophanase. The presence of tryptophanase is a significant distinction between *Escherichia coli* which produces it and is therefore indole positive, and the other enteric organisms. If indole is present, it reacts with Kovac's reagent to produce a red color. The reagent does not mix with the aqueous medium and forms a layer on the surface. The red color develops in that layer.

HYDROGEN SULFIDE PRODUCTION

Hydrogen sulfide (H_2S) is a by-product of the breakdown of cysteine by bacteria who produce the enzyme cysteine desulfurase. This enables the organisms to utilize the sulfur-containing amino acids for energy metabolism. The hydrogen sulfide is excreted and can be tested for by taking advantage of the formation of a black precipitate when it reacts with either silver, iron, or lead in the medium. To test for hydrogen sulfide, you must use a medium containing the sulfur-containing amino acids and a source of one of the metals mentioned above. Iron (as ferrous sulfate) or lead (as lead acetate) are the most commonly used sources. The medium is inoculated as a stab into an agar deep. When the hydrogen sulfide is produced, a black precipitate forms along the line of inoculation.

UREA HYDROLYSIS

Urea is produced as a by-product of protein and nucleic acid decomposition. If an organism produces the enzyme urease, it enables bacteria to detoxify urea, a waste product, and to release usable energy as it hydrolyzes it to ammonia and carbon dioxide.

$$O = C \begin{matrix} / NH_2 \\ \backslash NH_2 \end{matrix} + H_2O \xrightarrow{\text{urease}} 2 NH_3 + CO_2$$
(urea) (ammonia)

The ammonia reacts with the water in the medium to produce ammonium hydroxide, causing the pH to rise. Urea medium contains phenol red indicator (what are its pH range, and colors?) in addition to urea, and has an initial pH of about 6.5 to 6.8. Consult Exercise 12 to determine how the medium will change color as the ammonia is produced.

NITRATE REDUCTION

Nitrate reduction is a characteristic shown by several organisms, with the production of nitrite frequently used in diagnostic protocol for gram-negative rods. Nitrate is utilized as the final electron acceptor in anaerobic respiration, reducing the nitrate to nitrite.

$$NO_3^- + 2 e^- + 2 H^+ \xrightarrow{\text{nitrate reductase}} NO_2^- + H_2O$$
(nitrate) (nitrate)

Chemicals are used to detect the nitrite, as well as any remaining nitrate and other reduction products of the nitrate. After nitrite is produced, it may, in turn, be further reduced to several gaseous products,

including nitrogen gas, or to ammonia. The gaseous products are given off to the atmosphere, and the ammonia is frequently used for the synthesis of proteins, although it may be excreted into the medium under some circumstances. If an organism tests negative for the presence of nitrite, it can mean one of three things:

(1) the organism does not reduce nitrate;
(2) the organism reduced nitrate to nitrite, but also produces nitrite reductase which reduces the nitrite to ammonia; or
(3) denitrification of the nitrate has occurred, resulting in the reduction of nitrite to nitrogen gas.

If the nitrate test is negative for the presence of nitrite, an additional test for the presence of unreacted nitrate should be performed. Zinc dust is added to the tube. This will catalyze the reduction of nitrate to nitrite. If nitrate is still present in the media, it will be reduced to nitrite which will react with the reagents present to produce a red color.

LABORATORY OBJECTIVES

The objectives of this exercise are similar to those of Exercise 12 (Part I). Here, however, we can begin to look at the relationships between different pathways and functions within the cell. You should:

• Understand why some parts of amino acid molecules are excreted when the amino acids are used for energy metabolism.

• Obtain an appreciation for the metabolic capabilities of bacteria and how these capabilities can be used to identify bacteria.

• Become introduced to the characterization of bacteria on the basis of their phenotypic characteristics.

MATERIAL NEEDED
FOR THIS LABORATORY

A. HYDROLYSIS OF GELATIN

1. Twenty-four hour cultures to include:
   Enterobacter aerogenes
   Escherichia coli
   Proteus vulgaris

2. Nutrient gelatin (1 tube per organism)

B. INDOLE PRODUCTION

1. Twenty-four hour cultures to include:
Enterobacter aerogenes
Escherichia coli
Proteus vulgaris

2. Tryptone broth (1 tube per organism)

3. Kovac's reagent

C. HYDROGEN SULFIDE PRODUCTION

1. Twenty-four hour cultures to include:
Enterobacter aerogenes
Escherichia coli
Proteus vulgaris

2. Peptone iron agar (1 tube per organism)

D. UREA HYDROLYSIS

1. Twenty-four hour cultures to include:
Enterobacter aerogenes
Escherichia coli
Proteus vulgaris

2. Urea broth (1 tube per organism)

E. NITRATE REDUCTION

1. Twenty-four hour cultures to include:
Acinetobacter calcoaceticus
Escherichia coli
Pseudomonas aeruginosa

2. Nitrate broth (5 ml per tube)

3. Nitrate Reagents: Solutions A and B

4. Zinc dust

5. Toothpicks

LABORATORY PROCEDURE

A. HYDROLYSIS OF GELATIN

1. Obtain and label one tube of nutrient gelatin for each organism being tested.

2. Inoculate the nutrient gelatin tubes by stabbing with the inoculating needle into the center of the tube.

3. Incubate at 37°C for 24 – 48 hours.

4. Following incubation, place the nutrient gelatin cultures, along with an uninoculated control, in a cold-water or ice bath. After the control has

solidified, tilt each culture to determine if the gelatin has remained liquid. Any cultures that fail to solidify should be recorded as gelatinase positive.

5. Record your results on the Laboratory Report Form.

B. INDOLE PRODUCTION

1. Obtain and label one tube of tryptone broth for each organism being tested.

2. Inoculate the tubes with an inoculating loop or needle.

3. Incubate at 37°C for 24 – 48 hours.

4. Following incubation, add about 10 drops of Kovac's reagent to the tryptone broth. The reagent will form a layer on the surface and develop a red color if indole is present. The development of a red color should be recorded as positive for the production of indole.

5. Record your results on the Laboratory Report Form.

C. HYDROGEN SULFIDE PRODUCTION

1. Obtain and label one tube of peptone iron agar for each organism being tested.

2. Inoculate the tubes by stabbing with the inoculating needle into the center of the tube.

3. Incubate at 37°C for 24 – 48 hours.

4. Examine the peptone iron agar tubes for the presence of lead or iron sulfate. Any blackening of the medium indicates that hydrogen sulfide was produced.

5. Record your results on the Laboratory Report Form.

D. UREA HYDROLYSIS

1. Obtain and label one tube of urea broth for each organism being tested.

2. Inoculate the tubes with the inoculating loop or needle.

3. Incubate at 37°C for 24 – 48 hours.

4. Examine the tubes of urea broth. A red color indicates that the urea has been hydrolyzed and ammonia excreted, and should be recorded as a positive test.

5. Record your results on the Laboratory Report Form.

E. NITRATE REDUCTION

1. Obtain and label one tube of nitrate broth for each organism being tested.

2. Inoculate the tubes with an inoculating loop or needle.

3. Incubate at 37°C for 24 – 48 hours.

4. Add 3 drops of Nitrate Reagent A, then 3 drops of Nitrate Reagent B to each tube of nitrate broth. If nitrite is present, a distinctive red color will develop. The organisms is positive for nitrate reduction.

5. If a red color did not develop, add a small amount of zinc dust to the broth. If a red color develops following the addition of zinc, it indicates that nitrate, not nitrite, was present in the tube. The red color is due to the reduction of nitrate to nitrite by the zinc (a reducing agent). The test should be recorded as negative for nitrate reduction (the organism did not reduce the nitrate).

6. If no color develops following the addition of the zinc dust, then neither nitrate nor nitrite are present in the tube. The organism reduced the nitrate to nitrite and on to ammonia or free nitrogen. The test should be recorded as positive for both nitrate reduction and nitrite reduction.

7. Record your results on the Laboratory Report Form.

LABORATORY REPORT FORM

EXERCISE 13
PHYSIOLOGICAL REACTIONS OF BACTERIA, PART II

What was the purpose of this exercise?

Record the results you got for the following tests and organisms:

ORGANISM	HYDROLYSIS OF GELATIN	INDOLE PRODUCTION	HYDROGEN SULFIDE PRODUCTION	UREA HYDROLYSIS	NITRATE REDUCTION
Acinetobacter calcoaceticus					
Enterobacter aerogenes					
Escherichia coli					
Proteus vulgaris					
Pseudomonas aeruginosa					

QUESTIONS

1. What is hydrolysis? List three examples.

2. Why did you stab the peptone iron agar instead of inoculate only the surface?

3. Your microorganism grows well in the peptone iron agar, but does not form a black precipitate. What does this mean?

4. Why is gelatin less suitable than agar as a solidifying agent for bacteriological media?

5. When changing a baby's wet diapers, why does one smell ammonia?

6. In the nitrate reduction test, what does the presence of gas indicate?

7. Is nitrate reduction beneficial or harmful to farmers?

PHYSIOLOGICAL CHARACTERISTICS OF BACTERIA, PART III:
MISCELLANEOUS REACTIONS

BACKGROUND

Many bacteria produce enzymes that are secreted into the medium. For example, the hydrolysis of starch (Exercise 12) and the liquefaction of gelatin (Exercise 13) occur because some bacteria secrete hydrolytic enzymes that can depolymerize (break down into smaller units) these molecules. Many of the secreted enzymes have ecological or medical significance, and a few have diagnostic significance as well.

The clinical significance of some of these enzymes lies in their ability to attack substrates in host cell membranes, tissues, or body fluids, thereby interfering with host defense mechanisms or homeostasis. In other instances the enzymes are antigenic and can be tested for by serological reactions that provide evidence of exposure to the bacteria that produced the enzymes.

The enzymatic reactions we will examine in this exercise are ones that have been shown to be useful in the identification of unknown isolates. They are, of course, also important for the normal metabolism of the cell.

When bacteria produce flagella they are able to swim through liquids and soft media. They are said to be motile. Motility, as we will test for it here, is the result of the activity of flagella by those species of bacteria that produce them.

It is usually possible to test for enzymatic reactions by adding the enzyme's substrate to the bacterial culture and then testing for the appearance of the product or the disappearance of the substrate. For example, when you tested for starch hydrolysis in

Exercise 12, you grew the bacteria on starch agar and then added iodine to test for the hydrolysis of the starch.

SPECIFIC ENZYMATIC AND MOTILITY TESTS
Catalase Test

Catalase is an enzyme produced by many organisms, including man. Indeed, it is so commonly produced that those organisms that do not produce it are rare enough so that the lack of catalase is a significant diagnostic characteristic. The enzyme converts hydrogen peroxide into water and oxygen and does so vigorously enough to cause foaming. (What happens when you use hydrogen peroxide to cleanse a cut?)

$$2 \ H_2O_2 \ \rightarrow \ 2 \ H_2O \ + \ O_2$$
(hydrogen peroxide)

Hydrogen peroxide is produced when some organic molecules are directly oxidized. It must be rapidly removed because it is highly toxic to most cells—hence, most cells that are aerobic or facultative produce catalase, and those cells that do not do so are either obligate anaerobes or microaerophilic. Catalase is tested for by dropping one drop of hydrogen peroxide (the substrate) on a bacterial colony. If bubbles appear (the product), the organism is catalase positive. The catalase test is a very important one for the differentiating between the genera *Staphylococcus* and *Streptococcus* as well as between the genera *Bacillis* and *Clostridium*.

Oxidase Test

Oxidase is an enzyme involved in certain oxidation reactions in aerobic bacteria. A relatively few genera, notably *Neisseria, Pseudomonas,* and a few other gram-negative rods, produce it, making it an important diagnostic test. To perform the test, it is necessary to combine the oxidase reagent (the substrate) with the colony being tested. If the colony is positive, it will turn first pink, then purple, and eventually black (the colored products are the result of the enzymatic activity). It is sometimes more convenient to simply flood the plate with the reagent and observe the color changes. The oxidase reagent is, however, toxic, and positive colonies must be rapidly transferred if they need to be subcultured.

Coagulase Test

Coagulase is an enzyme produced by *Staphylococcus aureus,* which converts fibrinogen to fibrin blood plasma resuiting in the coagulation of blood plasma. When coagulase is produced by a pathogenic organism, it results in the formation of a fibrin meshwork that serves to protect the organism from the host's natural defense mechanisms. The enzyme is not produced by *S. epidermidis* or by any other gram-positive coccus. This test is very important because it allows for rapid identification of an important and commonly encountered pathogenic bacterium. A rapid screening test for coagulase may be completed by mixing a small amount of culture material with a drop of human or rabbit plasma. Try to emulsify the suspension. If the cells appear to agglutinate (clotting the plasma), the test is positive. (What are the substrate and product for this enzyme?) Cells that are coagulase negative will produce an even, nonagglutinated suspension. A more definitive and sensitive test requires incubation of the bacterium in plasma at 37°C. Coagulation within 24 hours is considered positive for the presence of a virulent strain of *Staphylococcus aureus.*

DNase Test

DNase is an excreted enzyme (like the gelatin and starch enzymes) that hydrolyzes DNA. A special medium, containing DNA and a dye, is used to measure this activity. The bacterial species to be tested is spottted on the surface of the plate and the culture incubated for 24 hours. If DNA is hydrolyzed (by DNase), the medium will change color. DNase production is an important diagnostic characteristic for the staphylococci and some gram-negative rods.

Esculim Hydrolysis

Esculin hydrolysis (and growth in a medium containing bile) is considered to be definitive for the Enterococci, a group of streptococci that are frequently encountered in clinical samples. This test uses a medium containing both esculin and bile. Positive organisms (Enterococci) will grow and turn the medium black. The black color is due to a hydrolysis product of esculin which reacts with the iron in the medium. Bile is added as an additional test for the enterococci. It has no effect on the hydrolysis of esculin, except that it limits the type of organism that can grow on the medium (it must be resistant to bile).

Motility

Whereas motility is not due to a chemical reaction, it is a characteristic of some organisms which can be readily observed by the use of appropriate media. Bacteria that have flagella are able to swim through liquids and soft media (less than one half the amount of agar that is usually found in plated or slanted media). This ability can be observed directly with a wet mount slide, or indirectly by using soft agar deeps. (Wet mounts are discussed in Exercise 2.) Soft agar deeps are inoculated with a single stab down the center of the deep. Motile bacteria will move outward from the stab, resulting in a medium with a very hazy and cloudy appearance. Nonmotile bacteria cannot move and therefore grow only along the stab, producing a clear and sharp stab line. Some media contain a dye that is turned red by bacterial action, making the above observations much more distinct. When you examine your results, the difference will be clear and obvious.

LABORATORY OBJECTIVES

The enzymes secreted by bacteria have both ecological and clinical significance. The production of these enzymes may also be of diagnostic significance. You should:

- Try to understand the ecological significance of excreted enzymes. For example, how do bacteria and other microbes cause the decomposition of dead plant and animal remains?

- Understand the possible clinical significance of certain bacterial enzymes (hemolysin, coagulase, collagenase, etc.). Those enzymes that are clinically significant may attack certain cells or inhibit certain processes in the host, or they may be serologically important.

- Appreciate the role that certain enzymes play in the metabolism and in the identification of bacteria.

MATERIALS NEEDED FOR THIS LABORATORY

FOR PROCEDURES A - C

1. Twenty-four hour cultures including:
   Lactococcus lactis
   Micrococcus luteus
   Pseudomonas aeruginosa
   Staphylococcus aureus
   Staphylococcus epidermidis

2. One nutrient agar plate per organism

A. CATALASE TEST

1. Hydrogen peroxide

2. Microscope slide

3. Pasteur pipette

4. Toothpick

B. OXIDASE TEST

1. Oxidase reagent

2. Filter paper

3. Petri plate

4. Toothpick

C. COAGULASE TEST

1. Coagulase Plasma

2. Sterile test tube

3. Pipettes

D. DNase TEST

1. DNase agar containing indicator—1 plate for every two organisms to be tested

2. Twenty-four hour cultures to include:
   Serratia marcescens
   Enterobacter aerogenes

E. ESCULIN HYDROLYSIS

1. Bile-esculin agar talls -- 1 tubefor each organism to be tested

2. Twenty-four hour cultures to include:
   Enterococcus faecalis
   Streptococcus mitis

F. MOTILITY

1. Motility agar tall—1 tube for each organism to be tested

2. Twenty-four hour cultures to include:
   Enterobacter aerogenes
   Staphylococcus epidermidis

LABORATORY PROCEDURE

FOR PROCEDURES A - C

1. Prepare a streak plate (see Exercise 8) of each organism.

2. Incubate at 37°C for 24 - 48 hours .

3. Following incubation, perform the following tests on the organisms:

A. CATALASE TEST

1. Spot-catalase Test: Place a drop of hydrogen peroxide on a clean slide. Using a toothpick, carefully remove an isolated colony from the streak plate and add it to the drop of hydrogen peroxide. Observe for bubbles. **NOTE:** Be sure to dispose of the toothpick as indicated by your instructor.
   OR
   Plate-catalase Test: Carefully add a drop of hydrogen peroxide to an isolated colony on the streak plate. Observe for bubbles.

2. If bubbles are present, the organism is catalase positive

3. Record your results on the Laboratory Report Form.

B. OXIDASE TEST

1. Spot-oxidase Test: Place a piece of filter paper into an empty petri dish. Place a few drops of oxidase reagent on the filter paper. Using a toothpick, transfer an isolated colony from the streak plate to the moistened filter paper. Observe for a color change.

OR

Plate-oxidase Test: Place one drop of the reagent on the colony to be tested. Observe for a color change.

2. A color change to pink, then purple indicates that the organism is oxidase positive.

3. Record your results on the Laboratory Report Form.

C. COAGULASE TEST

1. Slide screening test: Place two drops of coagulase plasma reagent in the center of a slide. Using a toothpick, transfer an isolated colony from the streak plate and emulsify it in the drop of plasma. If the organism is coagulase positive, it will agglutinate, whereas a coagulase negative organism will emulsify easily and produce a uniformly hazy suspension. **NOTE:** Your instructor may demonstrate this procedure. If you perform this test, be sure to dispose of the toothpick as indicated by your instructor.
OR
Tube incubation method: Using a sterile pipet, transfer 0.1 ml of coagulase plasma into a test tube. Add 0.4 ml of distilled water. Using your inoculating loop, transfer several colonies from your streak plate to the tube. Incubate at least four hours. A coagulase-positive organism will coagulate the entire volume of plasma within 24 hours.

2. Record your results on the Laboratory Report Form.

D. DNase TEST

1. Using your marker, divide the DNase plate(s) in half. Draw a circle the size of a dime in the center of each half. Inoculate one organism on each half of the plate, spreading it out to fill the circle.

2. Incubate at 37°C for 24 – 48 hours.

3. If the organism has produced DNase, there will be decolorization of the dye surrounding the area of growth. This would be a positive test for DNase production.

E. ESCULIN TEST

1. Using your inoculating needle, inoculate each tube of bile esculin tall with a single stab into the center of the tube.

2. Incubate the tubes at 37°C for 24 – 48 hours.

3. Observe the tubes. The presence of a black color along the line of inoculation indicates that esculin has been hydrolyzed. This reaction would be positive for the hydrolysis of esculin. If there is no black coloration produced, note whether the organism was able to grow in the presence of bile .

4. Record your results on the Laboratory Report Form. Use (+) to indicate a positive esculin test; (-) to indicate a negative esculin test; and (O) to indicate the absence of any growth .

F. MOTILITY

1. Inoculate the motility agar with a single stab into the center of the tube.

2. Incubate the tubes at 37°C for 24 – 48 hours.

3. Observe the growth along the stab line in the motility agar. If the organism is nonmotile, the stab will be clearly defined; if the organism is motile, the stab will be cloudy and hazy, indicating that the bacteria have moved through the medium.

4. Record your results on the Laboratory Report Form.

LABORATORY REPORT FORM

EXERCISE 14
PHYSIOLOGICAL REACTIONS OF BACTERIA, PART III

What was the purpose of this exercise?

Complete the following chart with your results:

ORGANISM	CATALASE	OXIDASE	COAGULASE	DNase Test	ESCULIN HYDROLYSIS	MOTILITY
Enterobacter aerogenes	▨	▨	▨		▨	
Enterococcus faecalis	▨	▨	▨	▨		▨
Lactococcus lactis				▨	▨	▨
Micrococcus luteus				▨	▨	▨
Pseudomonas aeruginosa				▨	▨	▨
Serratia marcescens	▨	▨	▨		▨	▨
Staphylococcus aureus				▨	▨	▨
Staphylococcus epidermidis				▨	▨	
Streptococcus mitis	▨	▨	▨	▨		▨

QUESTIONS

1. What is coagulase? How is it related to pathogenicity?

2. What is the medical significance of DNase, coagulase, and catalase?

3. Why don't obligate anaerobes require catalase?

4. A catalase test is performed on an isolated colony on a blood agar plate, and bubbles appear. The test is repeated by performing a spot-catalase test on a colony from the same pure culture. This time there are no bubbles. How can you explain the difference in results?

5. Using the tests performed in Exercise 14, devise a scheme to distinguish between *Enterococcus faecalis, Staphylococcus epidermidis,* and *Staphylococcus aureus.*

DEVELOPMENT AND USE OF DIAGNOSTIC KEYS

In the previous exercises, you have been learning about the varied enzyme systems that given organisms possess through an examination of the metabolic processes they perform and the waste products formed. When this information is combined with basic morphological and staining results it becomes possible to begin the identification process. In this exercise you will become familiar with the use of diagnostic keys and *Bergey's Manual of Determinative Bacteriology.*

BACKGROUND

Diagnostic keys are commonly used for the *identification* of organisms. The characteristics of an unknown microorganisms are compared to those of a wide variety of other organisms in an attempt to determine the genus and species identification of the unknown. It must be noted that this is **not** the same as *classification* of organisms. In classification, organisms are placed in taxonomic groups based on similarity of structures, where as identification places an organism into a known taxonomic group based on its characteristics. Classification is a complex process, often utilizing highly sophisticated testing procedures, such as DNA hybridization. Identification is a much simpler process which is not as concerned with the total characterization of the organism as it is with the utilization of a few easily determined test results which will enable the distinction between varied groups of microorganisms. Classification methods are most commonly employed in research laboratories where possible new microorganisms are being worked with. Identification methods are most commonly used in the clinical laboratory where rapid, accurate, and inexpensive methods must be used for microbe identification.

The major reference text for classification and identification has been *Bergey's Manual of Determinative Bacteriology.* As more and more information has become available for microorganisms, *Bergey's Manual* has become more of a reference for classification and nomenclature, and less useful as an identification tool. The 9th Edition of this text was retitled *Bergey's Manual of Systematic Bacteriology* when it was revised in the 1980s. This four-volume series discusses the classification and nomenclature of organisms, giving detailed descriptions of species and providing methods for enrichment and isolation. In 1994, *Bergey's Manual of Determinative Bacteriology* was published. It is not an abridged version, but rather was established to deal specifically with the identification of organisms. In addition, it contains changes in nomenclature that have occurred since the publication of *Bergey's Manual of Systematic Bacteriology.*

The classical approach to identification is the development of keys or diagnostic tables. Included are the use of dichotomous keys, profile analyses, and numerical identification. All of these methods utilize characteristics similar to those you have obtained in the previous exercises, including:

Cell morphology
Staining reactions

Colony morphology
Biochemical characteristics

THE DICHOTOMOUS KEY

A dichotomous key is a list of increasingly specific phenotypic traits that will allow you to assign a genus and species name to your organism by a step-wise elimination of other organisms. In this system, the identifying characteristics are used one at a time to eliminate unrelated organisms .

It is, in effect, a road map that takes you to the genus and species name of your unknown. It requires that you be able to define your isolate by answering simple, single-variable questions about it. Usually a dichotomous key's questions can be answered with a simple yes or no. For example, "Does the unknown ferment lactose?" or "Does the unknown hydrolyze gelatin?" In other cases, the question calls for a choice between two mutually exclusive conditions or states, such as "Is it gram-positive or gram-negative?" or "Is it a bacillus or a coccus?" When you put a dichotomous key into diagram form, it should look like a pyramid, the first test being at the top or apex, and the last tests and the names of the identified bacteria at the base.

In order to use a dichotomous key, you need only know which tests must be performed and then follow the key to determine the correct genus and species. Obviously, however, someone must have first determined which tests could be applied diagnostically and then determined the best sequence in which to use the tests. As you examine the sample key, note that the tests applied late in the key are much more specific than those used early in the key. The later tests usually distinguish single species, while the early ones typically distinguish genera or even groups of genera.

The disadvantages of dichotomous keys include the necessity to interpret the tests absolutely as either plus/minus (+/-) or yes/no. There is no accommodation of variant organisms, or those with atypical characteristics differing from the type strain. It is also true that a single error in test interpretation can result in misdiagnosis. Each dichotomous key is based on judgments about the proper sequence of tests used in the key. In a very real way, the dichotomous key is like a road map -there may be several routes to the same destination, but a wrong turn early in the route will leave you lost in a strange neighborhood.

AN EXAMPLE EXERCISE: DEVELOPMENT OF A DICHOTOMOUS KEY

A *bacterial characteristic data matrix* has been compiled using the same characteristics as those you have been determining in the previous exercises. That data will be used to write your own dichotomous key. A dichotomous key is often referred to as a flow chart for identification. In the example given here, all the necessary information is provided (See Table 15.1). You should try to understand that the three sample charts are simply three ways of presenting the same data.

The matrix you will be using is more extensive than the example shown here since you will be using data representative of all the tests performed in the previous exercises. You will use that data to develop your dichotomous key.

For the organism shown in the example, the first branch in the key might be the gram reaction and/or the shape of the cells. The bacilli can then be differentiated on the basis of the fermentation of lactose, and the cocci can be distinguished by either the coagulase reaction or the liquefication of gelatin.

The information contained in the Data Matrix Chart is presented below as it might appear in a flow diagram or as a "road map." Try to determine the relationship between the two ways of presenting these data.

The first branch in your key should be a characteristic that divides the group of organisms into two large, approximately equal groups. At each subsequent branch of your key, you should use increasingly more specific tests until individual organisms are identified. When you have completed the key, you should be able to use it to identify any of the bacteria that you worked with in the previous exercises (See Table 15.2).

There is yet another way of presenting the dichotomous key. It takes the form of a series of statements about the organism; it is probably the form that you will most likely encounter. The following table is a short dichotomous key using the examples in this exercise (See Table 15.3).

PHENOTYPIC PROFILE ANALYSIS

Profile analysis is a diagnostic procedure that simultaneously compares several tests to develop a profile of the unknown and compare it with profiles in a reference data bank.

Table 15.1		Data Matrix Chart—An Example			
NAME OF ORGANISM	GRAM REACTION	CELL SHAPE	FERMENT LACTOSE	LIQUIFIES GELATIN	COAGULASE TEST
E. coli	-	Bacillus	+	-	-
S. aureus	+	Coccus	+	+	+
S. epidermidis	+	Coccus	-	-	-
P. aeruginosa	-	Bacillus	-	+	-

The advantages of profile analysis include the ability to compare many tests at once, avoiding the problems created by "wrong turns," and the ability to accommodate variants that have one or more atypical characteristics. The disadvantage of this technique is that as additional characteristics are added to the profile, it is increasingly difficult to keep everything straight. Although the average human mind can reasonably handle up to ten or so traits, as many as 20 or 30 individual traits must often be used to complete a diagnosis. The development of computers has made profile analysis readily available; several commercial systems are on the market. Two of these systems will be considered in Exercise 17.

PHENOTYPIC PROFILE ANALYSIS— A BRIEF DISCUSSION

If we use the example given above, a profile analysis system would consider all five of the characteristics simultaneously. An organism that gave the results shown on the chart (the unknown) would be matched with *S. aureus*.

If one test varied from the above reference profile, the system would report the goodness of fits (e.g.,

"four/five" match) with *S. aureus*. In this case, however, since there are only two tests differentiating *S. aureus* from *S. epidermidis*, the profile analysis would alert you if the atypical test was one of the two needed to differentiate between these two bacteria.

LABORATORY OBJECTIVES

The requirements for this exercise are for you to become familiar with the use of phenotypic characteristics as tools for bacterial identification. To accomplish this, you will:

- Use phenotype characteristics in a data matrix chart to develop a diagnostic key that would allow anyone to identify any of the bacteria studied so far.

- Gain an understanding of how the identification of unknown bacterial isolates must follow a logical, step-by-step process.

- Gain an understanding of the differences, both advantageous and disadvantageous, between a dichotomous key and a phenotypic profile.

Table 15.2   Dichotomous Key—An Example

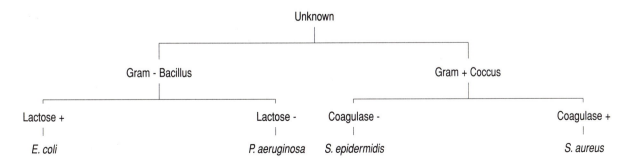

Table 15.3   The Dichotomous Key

1. Organism is a gram-negative bacillus
   a. Ferments lactose   *E. coli*
   b. Does not ferment lactose   *P. aeruginosa*
2. Organism is a gram-positive coccus
   a. Organism is coagulase positive   *S. aureus*
   b. Organism is coagulase megative   *S. epidermidis*

- Become familiar with the use of *Bergey's Manual of Determinative Bacteriology* for the identification of bacterial organisms.

LABORATORY PROCEDURE

1. Use the data matrix chart found in Table 15.5 to develop your own dichotomous key. Remember, there is more than one way that this key can be put together—and none of them is the "right" way!

2. You might consider simplifying your task by starting with certain assumptions:

   a. Can a separate key be used for major groups of bacteria, such as the gram-positive cocci, the gram-negative bacilli, and so on?
   b. Determine which reactions divide the largest number of organisms into manageable groups. For example, lactose fermentation, oxidase reaction, and hydrogen sulfide production might be used to separate the gram-negative bacilli into manageable groups, especially if the reactions are used sequentially.

   c. Use the approach that you are solving an unknown (as you will soon be doing). What is the most direct route to the name of the unknown?

3. Try using the key on two or three of your known organisms to see if it works. Better yet, have another student try it.

4. Make sure there is no ambiguity in the key. If a test is used more than once, consider if it can be moved up to a branch closer to the apex.

Table 15.4   Data Matrix Chart—An Example

NAME OF ORGANISM	GRAM REACTION	CELL SHAPE	FERMENT LACTOSE	LIQUIFIES GELATIN	COAGULASE TEST
E. coli	-	Bacillus	+	-	-
S. aureus	+	Coccus	+	+	+
S. epidermidis	+	Coccus	-	-	-
P. aeruginosa	-	Bacillus	-	+	-
Unknown	+	Coccus	+	+	+

Table 15.5   Data Matrix for Bacteria

ORGANISM	SHAPE	GRAM REACTION	ENDOSPORES	ACID-FAST	FLAGELLA	OXYGEN	GROWTH IN 10% NaCl	GLUCOSE FERMENTATION	SUCROSE FERMENTATION	LACTOSE FERMENTATION	MANNITOL FERMENTATION	STARCH HYDROLYSIS	METHYL RED TEST	VOGES-PRAUSKAUER	CITRATE	GELATIN LIQUIFICATION	INDOLE PRODUCTION	HYDROGEN SULFIDE	UREASE	NITRATE REDUCTION	CATALASE	OXIDASE	COAGULASE	DNase	ESCULIN HYDROLYSIS	MOTILITY
Staphylococcus aureus	C	+	-	-	-	F	+	+	+	+	+					+			w	+	+		+	+		-
Staphylococcus epidermidis	C	+	-	-	-	F	-	+	+	d	-					-			+	w	+		-	-		-
Lactococcus lactis	C	+	-	-	-	F	-	+		+	-										-	-				-
Moraxella (Branhamella) catarrhalis	C	-	-	-	-	A	-	-	-	-	-					-	-		-	+	+	+				-
Escherichia coli	B	-	-	-	+	F		+ G	d	+	+		+	-	-	-	+	-	-	+	+	-	-	-	d	+
Enterobacter aerogenes	B	-	-	-	+	F		+ G	+	+	+		-	+	+		-	-	-	+	+	-		-	+	+
Proteus vulgaris	B	-	-	-	+	F		+ G	+	-	-		+	-	-	+	+	+	+	+	+	-		+	d	+

ORGANISM	SHAPE	GRAM REACTION	ENDOSPORES	ACID-FAST	FLAGELLA	OXYGEN	GROWTH IN 10% NaCl	GLUCOSE FERMENTATION	SUCROSE FERMENTATION	LACTOSE FERMENTATION	MANNITOL FERMENTATION	STARCH HYDROLYSIS	METHYL RED TEST	VOGES-PRAUSKAUER	CITRATE	GELATIN LIQUIFICATION	INDOLE PRODUCTION	HYDROGEN SULFIDE	UREASE	NITRATE REDUCTION	CATALASE	OXIDASE	COAGULASE	DNase	ESCULIN HYDROLYSIS	MOTILITY
Bacillus subtilis	B	+	+	·	+	A		+				+									+					+
Mycobacterium smegmatis	B	+	·	+	·	A					+				+					+	+					·
Pseudomonas aeruginosa	B	·	·	·	+	A		+		·	+	·				+				+	+	+	·			+
Streptococcus pneumoniae	C	+	·	·	·	F	·			+	·			·							·					·
Clostridium sporogenes	B	+	+	·	+	AN															·	·				+
Acinetobacter calcoaceticus	B	·	·	·	·	A		+							+	·				+	+	·				·
Micrococcus luteus	C	+	·	·	·	A		·		·					·	+				·	+	+			·	·

KEY: B = bacillus; C = coccus; A = aerobe; An = anaerobe; F = facultative anaerobe; w = weakly positive; d = positive 11-89% of the time

LABORATORY REPORT FORM

EXERCISE 15
DEVELOPMENT AND USE OF DIAGNOSTIC KEYS

What was the purpose of this exercise?

Develop your diagnostic key. Remember that, as with any road map, there is more than one way to get to any destination.

QUESTIONS:

Use *Bergey's Manual of Determinative Bacteriology*, 9th Ed. to answer the following questions.

1. Where would you find a description of the Archaeobacteria?

2. In which of the four major category of bacteria are the facultatively anaerobic gram-negative rods found?

3. Where are prokaroytes described?

4. Find the pages where it gives a general approach to using the manual.

5. Where are the following **described?**

   Enterobacteriaceae

   Rickettsia

   Staphylococcus

   Treponema

6. What was the genus *Lactococcus* previously known as? What is its type species?

7. Where would you look to differentiate the endospore-forming bacteria?

USE OF DIAGNOSTIC KEYS:
BACTERIAL UNKNOWNS

One of the reasons you need to be familiar with the development of diagnostic keys is that you may have to apply that information to the solving of unknowns. In clinical microbiology, every specimen sent to the laboratory for analysis is an unknown. It is often necessary to identify nonclinical isolates because pathogenic bacteria are frequently found in environmental samples and in contaminated food and water. In short, when you send a culture from any source to the laboratory asking them to identify the predominant organism(s), you have presented them with an unknown.

BACKGROUND

It is not usually possible to identify bacteria simply on its morphology. Physiological characteristics provide a much more diverse set of criteria for identifying bacterial organisms. By combining both morphological and physiological characteristics, the identity of most organisms can be determined. S.T. Cowan and J. Liston in the 8th edition of *Bergey's Manual* propose the following steps to identifying an unknown:

1. Make sure that you have a pure culture.

2. Work from broad categories down to a smaller, specific category of organisms.

3. Use all the information available to you in order to narrow the range of possibilities.

4. Apply common sense at each step.

5. Use the minimum number of tests to make the identification.

6. Compare your isolate to type or reference strains of the pertinent taxon to make sure the identification scheme being used actually is valid for the conditions in your particular laboratory.

When difficulty is encountered in determining the identity of an unknown, you must check its purity, the appropriateness of the tests performed, the reliability of your techniques, and your correct use of the keys available. It should be noted that the most frequent mistakes in unknown identification is error in determining shape, gram reaction, and motility or is due to contamination of the original culture.

PROBLEM-SOLVING APPROACH TO UNKNOWNS

The first procedures used in any identification protocol must be those that isolate the organisms into pure cultures. The technician must be familiar with the bacteria likely to be encountered and be able to recognize those bacteria by their colonial morphology. This initial isolation is sometimes referred to as primary isolation.

After isolation is achieved, the identification of the unknown proceeds by an orderly and stepwise elimination of alternate choices. The first choices are on the basis of the cellular morphology and staining reactions of the isolated culture. At the very least, a gram stain is considered essential, followed by whichever other stains would prove helpful (e.g., acid-fast and metachromatic-granule stains).

Once the cell morphology, the staining reactions, and the colonial morphology are known, an educated guess often can be made as to the possible genus and species of the bacteria. Then, depending on the possibilities suggested by the information already available, specific biochemical tests are used to confirm the identity of the unknown or to determine the genus and species if it is not already known. Which biochemical tests do you use? That's where your knowledge of and experience with diagnostic keys will prove useful. A good diagnostic protocol will use the least number of tests to arrive at the solution. To do this, you must know which tests are important for the type of organisms you are trying to identify. For example, you would use a different set of tests for gram-positive cocci than you would use for gram-negative bacilli.

The diagnostic key that you developed in the previous exercise may be used to identify you unknowns or to quickly determine if the unknown you are working with is not one of the species included in the key.

PROCEDURE FOR SOLVING UNKNOWNS

First Step

Primary isolation: Streak plates for initial isolation and determination of colonial morphology.

Second Step

Staining reactions: Microscopic examination for determination of cellular morphology and staining reactions.

Third Step

Biochemical tests: Confirmation and identification by biochemical characterization, using a carefully determined set of tests.

The preferred way to solve an unknown is to carefully determine which tests should be used, and in what order. The worst way, and often the most con-

fusing way, to handle an unknown is to simply subject it to every available test, hoping that sufficient data will be generated to provide the correct answer. It is true that all of the needed data will have been generated, but because there was no careful stepwise elimination of alternative choices, the student will usually be confronted with what often seems to be hopeless confusion.

In clinical applications, the identification of the bacteria in submitted samples is usually expected within 72 hours, and often within 48 or less hours. Speed and accuracy are essential and can only be maintained if a planned protocol is followed. If you are working with a Gram-negative bacillus, you do not use tests that are useful for the identification of Gram-positive cocci. If a Gram-positive bacillus that looks like a diphtheroid is observed to produce metachromatic granules, you do not need to do the oxidase test, but you do need to use a few selected tests to confirm the identification of *Corynebacterium diphtheria*. On the other hand, if you isolated a gram-positive coccus, you would consider using the coagulase test, catalase test, mannitol fermentation, and gelatin liquefaction (in that order?).

In summary, the best approach to identifying a bacterial unknown is a problem-solving approach. After each test you must ask yourself "What possibilities are indicated by the results?" and "What tests are needed to further differentiate this isolate?" The surest way to become completely lost and confused is to attempt to apply tests in a shotgun fashion, hoping to hit upon the answer. Plan your approach carefully.

LABORATORY OBJECTIVES

You will be asked to identify unknown bacterial cultures. To successfully accomplish that task, you must:

- Apply all of the skills learned in the previous exercises, including isolation of pure cultures, microscopic examination of stained smears, and recognition of colonial morphoiogies.

- Use a problem-solving approach to develop a protocol for the orderly application of diagnostic tests and the interpretation of the results obtained from those tests.

- Learn to anticipate what is next when the results of one set of tests are known.

MATERIALS NEEDED FOR THIS LABORATORY EXERCISE

NOTE: The bacterial cultures include, but are not limited to, bacteria covered in the previous exercises.

1. Unknown culture: Your unknown specimen represents either a urine infection or a wound infection and contains two organisms.

2. TSA plates.

3. All routine staining reagents.

4. Media and chemical reagents as needed to identify the unknown organisms.

LABORATORY PROCEDURE

HINTS AND SUGGESTIONS

1. **Never** discard a culture until any subcultures of it have grown up.

2. **Always** make a streak plate of your isolated cultures (even if they were derived from a single, well-isolated colony).

3. **Always** keep a culture of your unknown isolates (a slant is preferred) until you are completely finished with them. You may need to refer back to them.

4. **Always** keep all slides of your unknowns until you are completely finished with them.

5. **Always** include controls in your tests. For example, use uninoculated tubes of the fermentation media to compare colors after incuba-

tion. Use an uninoculated tube of gelatin as a negative control.

6. **Always** keep detailed notes of your laboratory work. It is almost always necessary to go back over them to determine some previous test results or to see if the results were interpreted correctly.

7. **Always** keep calm.

IDENTIFICATION PROCEDURES

1. The protocol you will follow for this laboratory exercise will be one you develop yourself. You must decide upon the protocol as you acquire data about your unknowns. You must decide what to do next.

2. The following initial steps are recommended:

   a. Make gram stains and streak plates of your culture.

   b. On the basis of the gram stains, and while you are waiting for the streak plates to grow, plan what tests will be used. Consult your keys from Exercise 15.

3. Continue with the testing until you have identified each organism in your unknown culture.

4. Complete a Bacterial Characteristics Form for each organism isolated, filling in only those results used in the diagnosis. Include under NOTES the sequence of tests performed as well as the reason for choosing which test to do next.

Exercise Sixteen • *Use of Diagnostic Keys*

NOTES

LABORATORY REPORT FORM

EXERCISE 16
USE OF DIAGNOSTIC KEYS

What was the purpose of this exercise?

RESULTS:

Unknown Number _____   Source _____

Organism #1:

Cellular Morphology and Staining Reactions:

Cell shape: _____

Gram stain reaction: _____

Endospore formation: _____

Acid-fast reaction: _____

Capsule formation: _____

Flagella: _____

Metachromatic granules: _____

Other morphological features: _____

Colonial Morphology:

Colony pigmentation: _____

Color of agar around colony: _____

Shape of colony: _____

Colony margin: _____

Carbohydrate Reactions:

    Glucose fermentation: _____

    Lactose fermentation: _____

    Sucrose fermentation: _____

    Mannitol fermentation: _____

    Hydrolysis of starch: _____

    Methyl red test: _____

    Voges Proskauer test: _____

    Utilization of citrate: _____

    Thioglycollate growth: _____
        (Aerobic/anaerobic)

Protein and Other Reactions:

    Hydrolysis of gelatin: _____

    Indole production: _____

    Hydrogen sulfide production: _____

    Catalase reaction: _____

    Oxidase reaction: _____

    Coagulase test: _____

    Urea hydrolysis: _____

    Nitrate reduction: _____

    DNase activity: _____

    Motility: _____

IDENTITY OF UNKNOWN #1   _____

NOTES: Include the sequence of tests performed (your protocol) as well as the basis for deciding which test to perform next.

Organism #2:

Cellular Morphology and Staining Reactions:

        Cell shape: _____

        Gram stain reaction: _____

        Endospore formation: _____

        Acid-fast reaction: _____

        Capsule formation: _____

        Flagella: _____

        Metachromatic granules: _____

        Other morphological features: _____

Colonial Morphology:

        Colony pigmentation: _____

        Color of agar around colony: _____

        Shape of colony: _____

        Colony margin: _____

Carbohydrate Reactions:

        Glucose fermentation: _____

        Lactose fermentation: _____

        Sucrose fermentation: _____

        Mannitol fermentation: _____

        Hydrolysis of starch: _____

        Methyl red test: _____

        Voges Proskauer test: _____

        Utilization of citrate: _____

        Thioglycollate growth: _____
                (Aerobic/anaerobic)

Protein and Other Reactions:

        Hydrolysis of gelatin: _____

        Indole production: _____

        Hydrogen sulfide production: _____

        Catalase reaction: _____

        Oxidase reaction: _____

        Coagulase test: _____

        Urea hydrolysis: _____

        Nitrate reduction: _____

        DNase activity: _____

        Motility: _____

IDENTITY OF UNKNOWN #2 _____

NOTES: Include the sequence of tests performed (your protocol) as well as the basis for deciding which test to perform next.

QUESTIONS:

1. What was the basis of your final determination of the identity of:

    Unknown # 1?

    Unknown #2?

2. What would have been the effect if there was an error in any of the test results.

RAPID DIAGNOSTIC TESTS:
PROFILE ANALYSIS

In Exercise 15 you developed a dichotomous key for the identification of bacterial isolates. You should recall that one of the problems with such keys is that they do not allow for consideration of variant strains. Another problem is the time involved. The clinical microbiology lab must be able to determine the identity of an unknown organism as quickly as possible.

Profile Analysis identification, by estimating the most likely match, rather than an exact match, does allow consideration of variant strains. In addition, several profile analysis systems have been developed which enable the rapid identification of unknowns.

BACKGROUND

A profile analysis uses the results of a selected series of tests to compare your isolate with a 'library' of organisms whose biochemical characteristics are known. All of the tests receive approximately equal weighting and, therefore, no single test result can significantly alter the overall profile. Since profile analysis requires only a "best fit," rather than an exact fit, a 100 % agreement between the unknown and the known is not required. Of course, if a 100% fit is possible, all the better. Most profile analyses report the degree of "fit" and reject those with less than some predetermined level (such as 90%).

Comparing the profile of the unknown with those of the organisms in the 'library' can be a tedious job. Imagine searching through a long table of (+)s and (-) s until you found one that matched. Fortunately such a tedious search is not necessary. Most profile analysis systems can be easily adapted to a computer analysis or to a relatively simple mathematical coding process that greatly simplifies the comparison process.

In developing rapid diagnostic test systems, accuracy and reliability must not be sacrificed. One factor that has helped maintain accuracy is the elimination of any culturing beyond the initial isolation of the organism. This eliminates much of the potential for contamination.

MULTIPLE AND RAPID TEST SYSTEMS

Two major approaches have appeared to aid in providing not only rapid but accurate diagnosis of organisms. The first is the utilization of multiple test systems. These incorporate a number of media into a single unit enabling the performance of multiple inoculations in a very short period of time. Incubation of 24 – 48 hours, however, is often still required. The other major approach is that of "rapid" tests—systems that provide reliable results in 2 to 4 hours. Because this is not long enough for significant growth to occur, these systems use varied substrates that can utilize preformed or constitutive enzymes always present in the organisms being tested.

The advantages of these systems include: (1) their short incubation time; (2) direct inoculation from an isolated colony; (3) ease of handling; (4) cost-effectiveness; (5) minimal storage space; (6) ease of adjustment to taxonomic change; and (7) ready tie in with computer services or other data bases.

There are, however, several disadvantages which must be addressed. These include: (1) varied accuracy dependent upon skill of the user; (2) difficulty in accurate color interpretation due to weak reactions; (3) possibility of media carryover between compartments; (4) inadequacy of test system with some organisms; and (5) occasional need for additional testing. To minimize the potential problems, it is important that the user carefully follow the manufacturer's directions as to the preparation of the inoculum, the incubation times, and the test interpretations.

TYPES OF SYSTEMS

A number of modifications of conventional testing systems have been developed to facilitate the rapidity of bacterial identification. These include the more accurate viewing of organisms, use of smaller and more rapid inoculations, a decrease in incubation time, the automation of the entire process, and the utilization of more systematized identification systems.

Modified Conventional Systems

One of the quickest ways to determine the identity of an organism is microscopic examination. As we have seen, however, it is very difficult to determine the identity of an organism microscopically. If we add an immunological reagent and marker system it is possible to quickly and accurately identify microorganisms. Examples include the florescent staining systems for *Chlamydia, Legionella, Neisseria gonorrhoeae, Treponema pallidum,* and several fungal and parasitic organisms.

Another rapid system is the utilization of various agglutination tests. These tend to be less sensitive than the conventional identification methods and are best used as screening systems. Examples include the rapid screening systems for the hemolytic streptococci, *Staphylococcus aureus,* several of the organisms which cause meningitis, *Legionella,* and *Clostridium difficile.*

Conventional testing systems have also been modified to provide more rapid results. Small volumes of media are inoculated with heavy suspensions of the organism to be identified and paper disks are impregnated with test reagents to provide rapid test determination. Tests that have been modified in this manner include the nitrate reduction, urease production, and decarboxylations.

Multitest Systems

Multitest systems are conventional tests packaged in such a manner as to enable a single inoculation of several media or the easy inoculation of several media packaged in small volumes. These provide specific patterns of test results which can be easily compared with the expected results of known organisms. Many of these systems require overnight incubation, but some will give reliable results in as little as 4 to 6 hours. Two of the most commonly used systems for the identification of enteric pathogens is the Enterotube II (Becton-Dickinson Microbiology Systems, Inc.) and API 20E (Analytab Products, Inc.) will be discussed in detail when we discuss their use in this exercise.

Some of the more recently developed test systems have been adapted so as to not require overnight incubation. These depend upon the detection of preformed enzymes present in a heavy suspension of the organism, giving results following a 4 hour incubation. The Micro-lD system uses several of these enzyme detection systems. Additional 4 hour systems utilize chromogenic substrates which give more rapid color changes and therefor a rapid reading is possible. An example of this type of system is the RapID STR for identification of streptococci and RapID SS/U for urinary tract bacteria.

Automated Systems

A number of automated identification systems have been developed in recent years. These utilize growth or changes in pH indicators to indicate reactions. Following an incubation period of 3 to 6 hours, light-scatter photometry is used to determine test results and the identity of the organism. These methods are rapid and accurate, but their cost prohibits their use in many laboratories.

ENTEROTUBE II MULTITEST SYSTEM

The Enterotube II consists of a single tube with 12 compartments, each containing a different medium, and a self-contained inoculating needle. It was developed to distinguish among the various Enterobacteriaceae. The needle is touched on an isolated colony and drawn though the tube to inoculate the media. Following incubation, fifteen biochemical determinations can be made based on colorimetric changes. The varied positive and negative test results are converted into a numerical ID value which is either looked up in a code book or determined by the ENCISE computer identification system. The tests included in the Enterotube II system and possible results are shown in Table 17.1. This was one of the first multitest systems developed and is still one of the easiest to inoculate and read.

Table 17.1   Enterotube Biochemical Reactions.

TEST	INITIAL COLOR	FINAL COLOR	COMMENTS
GLU	Red	Yellow	Acid from glucose
GAS			Gas from glucose
LYS	Yellow	Purple	Lysine decarboxylase
ORN	Yellow	Purple	Ornithine decarboxylase
H₂S		Black precipitate	Ferrous ions react with sulfide ions, forming black precipitate
IND	Beige	Red	Kovacs' reagent added to H2S/IND compartment for detection
ADON	Red	Yellow	Adonitol fermentation
LAC	Red	Yellow	Lactose fermentation
ARAB	Red	Yellow	Arabinose fermentation
SORB	Red	Yellow	Sorbitol fermentation
VP	Beige	Red	Voges-Proskauer reagents detect acetoin
DUL	Green	Yellow	Dulcitol fermentation
PA		Black precipitate	Phenylpyruvic acid released from phenylalanine combines with iron salts
UREA	Yellow	Pink	Ammonia changes pH of medium
CIT	Green	Blue	Citric acid used as carbon source

API 20E SYSTEM

The API 20E system consists of twenty miniature cupules on a plastic strip for the identification of the Gram-negative enteric bacteria. In the bottom of each is a dehydrated medium. These are reconstituted when the inoculum is added. Anaerobic conditions can be maintained by layering with mineral oil, while aerobic conditions are maintained by completely filling the cupule with inoculum. The various tests and results for the API 20E are shown in Table 17.2. Following an incubation of 18 to 24 hours, 20 test results are read directly with an additional four able to be read following the addition of reagents. These results are used to determine a code number which can be identified using the code book or by means of a computer.

LABORATORY OBJECTIVES

In this exercise phenotypic characteristics of bacterial isolates are used to develop a phenotypic 'profile' of the unknown. The profile thus developed is matched with a list of known profiles to identify the unknown. To understand the principles demonstrated here, you should:

- Compare the 'profile' method used in this exercise with the 'dichotomous key' method used in Exercise 18.

- Compare the utilization of the Enterotube II and API 20E systems for the identification of enteric pathogens.

- Understand why a single phenotypic characteristic will have a much greater impact in a dichotomous key than it would in a profile analysis key.

MATERIALS NEEDED FOR THIS LABORATORY

You may be performing both or only one of the rapid test procedures as indicated by your instructor.

1. Twenty-four to 48 hour streak plate of one of the following organisms. The colonies should be well isolated.
   Escherichia coli
   Enterobacter aerogenes
   Proteus vulgaris
   Proteus mirabilis
   Salmonella typhimurium *

Table 17.2   API 20E Biochemical Reactions.

TEST	INITIAL COLOR	FINAL COLOR	COMMENTS
ONPG	Yellow	Colorless	Lactose fermentation
ADH	Red	Yellow	Arginine dihydrolase
LDC	Red	Yellow	Lysine decarboxylase
ODC	Red	Yellow	Ornithine decarboxylase
CIT	Dark blue	Green	Citrate utilixation
H$_2$S	Black	None	Hydrogen sulfide production
URE	Red	Yellow	Urea hydrolysis
TDA	Brown	Yellow	Tryptophane deaminase
IND	Red ring	Yellow	Indole production
VP	Red	Colorless	Acetoin detected
GEL	Diffusion	None	Gelatin liquification
GLU	Yellow	Blue	Glucose fermentation Bubbles indicate the reduction of nitrate to N$_2$
MAN	Yellow	Blue	Mannitol fermentation; Detection of catalase
CAT	Bubbles		
INO	Yellow	Blue	Inositol fermentation
SOR	Yellow	Blue	Sorbitol fermentation
RHA	Yellow	Blue	Rhamnose fermentation
SAC	Yellow	Blue	Saccharose fermentation
MEL	Yellow	Blue	Melibiose fermentation
AMY	Yellow	Blue	Amygdalin fermentation
ARA	Yellow	Blue	Arabinose fermentation
OXI	Purple		Oxidase activity
MOT			Motility

Shigella, sp.*
Vibrio parahemolyticus *
* indicates pathogenic organisms. They may be presented as demonstrations.

A.   ENTEROTUBE II SYSTEM

1. Enterotube II

2. Test reagents:
   Kovac's reagent
   20% KOH containing 5% alpha-naphthol

3. Syringe and needle (optional)

4. Identification code manual for Enterotube II

B.   API 20E SYSTEM

1. Screw top tube containing about seven (7) ml of sterile saline (0.9%) for each organism being identified.

2. API 20E incubation card, with individual plastic incubation chamber and cover

3. Sterile transfer pipettes.

4. Sterile mineral oil

5. One tube of sterile water (5.0 ml).

6. Test reagents:
   Kovac's reagent
   10% ferric chloride
   40% potassium hydroxide
   6% -naphthol
   Nitrate reagent A
   Nitrate reagent B
   Oxidase reagent
   1.5% hydrogen peroxide

LABORATORY PROCEDURES

A.   ENTEROTUBE II SYSTEM (See Figure 17.1)

1. Label the Enterotube II with your name and the culture identification. Obtain a streak plate of the organism to be tested.

2. Remove both caps from the Enterotube II. One end of the wire is straight and is used to pick up the inoculum; the bent end of the wire is the handle. Holding the Enterotube II, touch the needle end to an isolated colony. Be careful not to touch the agar with the needle.

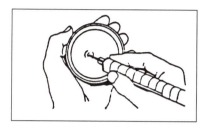

Pick: A well isolated colony directly with the tip of the Enterotube® II inoculating wire. A visible inoculum should be seen at the tip and the side of the wire. Avoid touching agar with wire. Utilize one or more Enterotube® II to pick additional colonies as experience dictates.

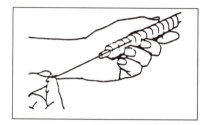

Inoculate: Enterotube® II by first twisting wire, then withdrawing wire through all twelve compartments using a turning motion.

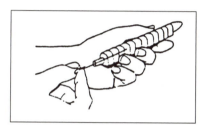

Reinsert wire (without sterilizing) into Enterotube® II, using a turning motion through all 12 compartments, until the notch on the wire is aligned with the opening of the tube. The tip of the wire should be seen in the citrate compartment. *Break wire* at notch by bending.

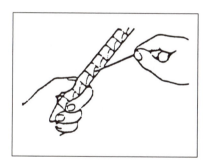

Punch holes with the broken off part of the wire through the foil covering the air inlets of the last eight compartments (adonitol, lactose, arabinose, sorbitol, Voges-Proskauer, dulcitol/PA, urea and citrate) in order to support aerobic growth in these compartments. Replace both caps. Incubate at 35 to 37°C for 18 to 24 h with Enterotube® II lying on its flat surface. Separate each Enterotube® II slightly to allow for sufficient air circulation.

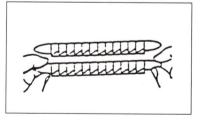

Interpret and record all reactions with exception of indole and Voges-Proskauer. (For complete instructions on how to read results of Enterotube® II see Results section.) All other tests must be read before the indole and Voges-Proskauer tests are performed as these may alter the remainder of the Enterotube® II reactions.

Figure 17.1   Inoculation of the Enterotube.

3. Inoculate the Enterotube II by holding the bent end of the wire and twisting. Draw the wire through all twelve compartments using a rotating motion. DO NOT remove the wire from the last compartment.

4. Re-insert the needle through the compartment until the notch on the wire is aligned with the opening of the tube (the tip should be in the citrate compartment). Break the needle at the notch by bending it. The portion of the needle remaining in the tube will maintain the anaerobic conditions necessary for fermentation, gas production, and decarboxylation.

5. Punch holes with the broken-off wire through the foil covering the air inlets of the last eight compartments (adonitol through citrate) to provide aerobic conditions. Replace the caps on both ends of the tube.

6. Incubate the tube by placing it on its flat surface in a 37°C incubator for 24 hours.

7. Read and record all of the reactions on the Laboratory Report Form. Be sure to record all results **before** performing the indole and V-P tests. Indicate each positive reaction by circling the number appearing below the appropriate compartment of the Enterotube 11 outlined in the Laboratory Report.
   Indole test: Make a small hole in the plastic film covering the H_2S/Indole compartment. Add 1 to 2 drops of Kovac's reagent. An alternative method would be to inject the Kovac's reagent using a syringe and needle. A positive test is indicated by a red color within 10 seconds.
   V-P test: Add 2 drops of 20% KOH containing 5% alpha-naphthol to the V-P compartment. A positive test is indicated by development of a red color within 20 minutes.

8. Add the circled numbers within each bracketed section. Enter the sum in the space provided. Note that the V-P test is only used as a confirmatory test. Read the resultant five digit number and locate it in the code book.

9. Dispose of the Enterotube as indicated by your instructor.

B. API 20E SYSTEM (See Figure 17.2)

1. Label one tube of sterile saline and one API 20E system with your name and the culture identifi-cation. Obtain a streak plate of the organism to be tested.

2. Using your inoculating loop, transfer several colonies to the saline tube, using aseptic techniques. Tighten the cap of the tube and gently mix it to obtain an even suspension.

3. Use a sterile transfer pipette to inoculate each of the microtubules on the incubation card. Pay careful attention to the amount of fluid that must be added to each microtubule.
   NOTE: Avoid leaving air bubbles in the microtubules. The formation of the bubbles can be minimized by touching the tip of the pipette to the inner edge of the microtubule as you SLOWLY release a small amount of suspension. If there are large bubbles, they can be teased out with a sterile inoculating needle.
   MINERAL OIL OVERLAY: Fill the tube part of the ADH, LDC, ODC and URE microtubules (to the level of the opening) with the saline suspension, THEN add enough sterile mineral oil to COMPLETELY fill the microtubule.
   COMPLETELY FILL: The CIT, GEL, and VP microtubules should be completely filled with the saline suspension.
   FILL ONLY the tube part of ALL REMAINING microtubules.

4. Pipette about 5 ml of sterile water into the BOTTOM PART of the API 20E plastic incubation tray. Rock the chamber to distribute the water into the indentations molded in it for that purpose. Place the inoculated incubation card into the chamber, cover it with the chamber lid, and carefully place the chamber in an incubator.

5. Incubate the chamber for at least 18, but not more than 24, hour. If the results cannot be read after 24 hours, remove the chamber from the incubator and refrigerate.

6. If the glucose (GLU) microtubule is NEGATIVE, and less than three positive reactions are observed in the other tubules, ask your instructor for assistance. NOTE: The demonstration organisms used in this exercise should give positive results within the first 24 hours. If you are using other gram-negative bacilli you may encounter some bacteria that will require an

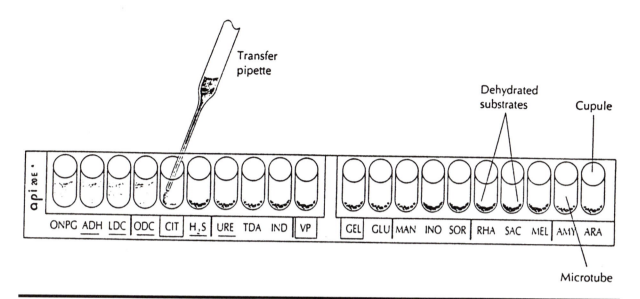

Figure 17.2   API 20E Test System.

additional incubation period. Specific instructions may be obtained from your instructor.

7. If ANY three tubules are POSITIVE, OR if glucose is POSITIVE, remove the chambers from the incubator and continue with the diagnostic tests. Add the necessary reagents for the following tests:

    Indole test: Add 1 drop of Kovac's reagent. A red ring after 2 minutes indicates a positive reaction.

    TDA test: Add 1 drop of 10% ferric chloride A positive test is brown-red. Indole-positive organisms may produce an orange color; this is a negative TDA reaction.

    V-P test: Add 1 drop of 40% KOH followed by 1 drop of 6% alpha-naphthol. A positive reaction produces a red color (not pale pink) after 10 minutes.

    Nitrate reduction: Before adding reagents, observe the GLU microtubule for the presence of bubbles. Bubbles indicate reduction of nitrate to nitrogen. Add to the GLU tubule 2 drops of nitrate reagent A (dimethyl- naphthylamine) and 2 drops of nitrate reagent B (sulfanilic acid). A positive reaction is the development of a red color in 23 minutes.

    Catalase test: After reading the results of the carbohydrate tests, add a drop of 1.5% hydrogen peroxide to the MAN microtubule.

8. Record all results on the Laboratory Report Form.

9. Determine the 7 digit identification number for your organism and determine its identity.

10. Dispose of the API 20E system as indicated by your instructor.

NOTES

Name _____

Section _____

LABORATORY REPORT FORM

EXERCISE 17
RAPID DIAGNOSTIC TESTS

What was the purpose of this exercise?

ENTEROTUBE® II

Circle the number corresponding to each positive reaction below the appropriate compartment. Then determine the five-digit code.

V-P results: _____

Identification of organism: _____

API 20E

Complete the following table and calculate the seven-digit code for the organism.
(1) Indicate if the test os positive or negative.
(2) Write the numerical value (1, 2, or 4) for each positive test.
(3) Add the values for each test in each group and write that number in the space provided.

API 20E REACTIONS

TEST	REACTION	VALUE	SCORE	TRIPLET SCORE
ONPG		1		
ADH		2		
LDC		4		
ODC		1		
CIT		2		
H2S		4		
URE		1		
TDA		2		
IND		4		
VP		1		
GEL		2		
GLU		4		
MAN		1		
INO		2		
SOR		4		
RHA		1		
SAC		2		
MEL		4		
AMY		1		
ARA		2		
OXI		4		

Seven-digit code   /   /

Identification of organism: _____

QUESTIONS

1. Did you get the same results with the Enterotube II and API 20E systems for your culture?

2. Which method, the Enterotube II or API 20E, did you prefer and why?

3. How is it possible for the same species to have 2 or more identification codes with either the Enterotube II or API 20E systems?

4. Why should the first digit in the Enterotube II code and the fourth digit in the API 20E code always be equal to or greater than 2?

5. The following results were gotten by two sets of students who performed API 20E tests on the same organisms. Which tests differed in their interpretation of the results?

    a) 3/752/677 vs. 3/652/677: _____

    b) 7/345/001 vs. 7/347/201: _____

    c) 3/432/654 vs. 2/432/654: _____

    d 5/362/135 vs. 5/362/137: _____

6. Compare the codes for the Examples (c) and (d) in the previous question. Explain why the difference in one of the pairs of codes, but not the other, would produce a significant problem when using a dichotomous key, but not necessarily in a profile key.

MICROBIAL POPULATION COUNTS:
MICROSCOPE COUNTING METHODS

In Exercise 4 we learned that the microscope may be used to measure the size of cells, and that once the length and/or diameter was determined, the volume, weight, and surface area of the cells could be calculated. In this exercise we will learn how to use the microscope to count the number of cells in a population.

There are two general methods of determining the size of cellular populations. One method relies on the actual counting of the cells using specially prepared microscope slides, counting chambers, or electronic particle counters. In the second method, the cells in the population are allowed to grow and the resultant colonies are counted or some other biological activity (such as increased protein, carbon dioxide production, or acid production) is measured. In Exercise 19 you will determine the size of a bacterial population by the plate count method, but in this exercise, you will use direct microscope counting methods to actually count yeast and bacterial cells

BACKGROUND

When more than one method is available to determine what appears to be the same data (e.g., the size of the population), you must often decide which of the methods is best. The answer to this question frequently lies in an understanding of the limitations of the methods and an understanding of what each of those methods actually measures. There certainly are many circumstances where plate counting will give all the information you need, but just as certainly, there are circumstances where microscopic counting

methods provide the quickest and best route to your data. You must choose the method that gives you the most useful data. Microscopic counting methods often provide information that is just not possible to obtain in any other way. Some examples:

- A microscope count includes all cells in the population, whereas a plate count includes only those cells that are capable of growing under the environmental conditions that were used in the experiment—medium, temperature, pH, oxygen content, and so on. For example, anaerobic bacteria will be unable to grow using any of the methods that are not strictly anaerobic. Similarly, if you were to use strict anaerobic methods, the aerobic organisms would be inhibited.

- A plate count cannot, of course, determine the number of nonviable cells in the suspension. For example, if you complete plate counts on a milk sample before and after pasteurization, you should get very different results. A comparison of the counts obtained by microscope and plate-count methodology would yield some information about the quality of the milk before pasteurization and about the effectiveness of the pasteurization method.

- A microscope count will sometimes be the only practical way of determining the size of the population or of measuring the number of cells in a given volume. A blood count is a good example.

DIRECT MICROSCOPE COUNTING PROCEDU RES

A.   COUNTING CHAMBERS

A counting chamber is a specially constructed glass slide that has a counting grid of known dimensions etched on its surface. A coverslip is supported a known distance above the etched surface. Examine Figure 18.1. Notice the location of the counting chamber and the counting grid.

Figure 18.2 shows, in a composite diagram, a counting chamber grid with a Neubauer ruling, and how it might look if you were counting yeast cells or bacteria. The square in the center of the grid, bounded by the double lines, is 1 mm². The volume of the fluid in the area bounded by the grid can be calculated because the dimensions of the grid and the space above it are known. Think of the counting area as a chamber bounded on the top by the coverslip, on the bottom by the chamber itself, and on the sides by the dimensions of the counting grid.

The center square, bound by the double lines, has an area of exactly 1 mm². This center square is subdivided into 25 smaller squares, each of which has sides 0.20 mm (or 200 µm) long. They are bounded by heavy lines and are 0.04 mm² (1/25 mm²) The smallest squares (bound by thin lines) have sides that are 0.05 mm (50 µm) long and which are 0.0025 mm² (1/400 mm²).

The grid to be used for counting will depend upon the number and size of the cells in the fluid being counted. For example, if you were counting bacteria, you would use the smaller squares (bounded by the thin single lines), but if you were counting yeast cells, the larger squares (bounded by the heavy lines) would be used. Different magnifications may also be used when counting the cells.

In practice, you would count the cells in several squares (at least five), calculate the average number of cells in each square, and then multiply by the appropriate numbers to obtain the number of cells per milliliter. You need to know the volume of the

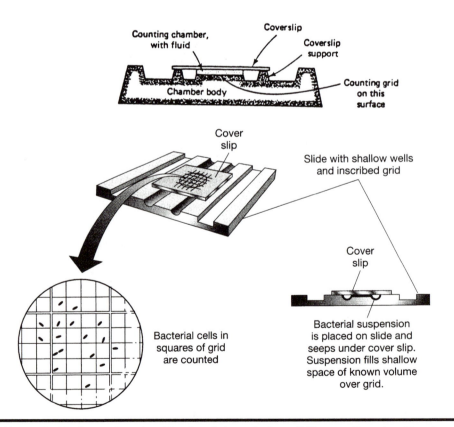

Figure 18.1   Above, the Neubauer Counting Chamber; below, the Petroff-Haussser Counting Chamber.

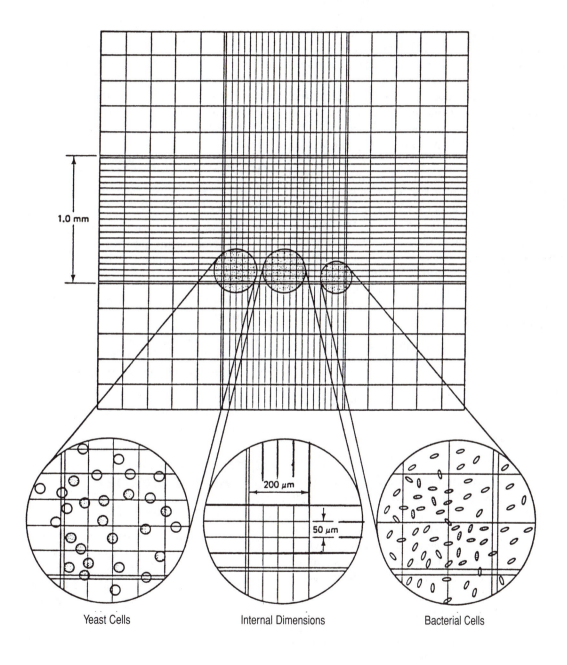

Figure 18.2   Cell-counting Chamber

fluid being counted in the chamber (multiply the area of the grid by the depth of fluid over the grid) and the dilution of the fluid used to fill the counting chamber.

Typically, the counting-chamber grid used to count bacteria or yeast cells is the central, 1 mm^2 grid. The depth of the space above the grid in a Neubauer chamber is 0.10 mm. The volume of a chamber with a depth of 0.10 mm would therefore be 0.10 mm^3 or 0.10 ml. With the Petroff-Hauser chamber, the depth is only 0.02 mm, giving a resultant volume of 0.02 mm^3 or 0.02 ml.

In making the actual count, you should ignore the cells touching the upper and left borders and count all the cells that touch the lower and right margins (even if most of the cell appears to lie on the wrong side of the line).

In Figure 18.2, there are 17 countable (shaded) yeast cells in the section of the counting grid shown (five are touching the upper and/or left margin). To determine the number of yeast cells in the sample, you would need to multiply the average number counted per square (in this case we'll use 17) by the number of squares in the grid (25). This will give you the number of cells in the volume of the counting chamber (0.10 mm^3 or 0.10 ml if you are using a Neubauer chamber).

17 cells/square X 25 squares/grid = 425 cells/0.10 ml

This would then need to be multiplied by 10 to determine the number of cells in 1 ml of the *diluted* sample.

425 cells/0.10 ml X 10 = 4250 cells/1.0 ml

To determine the total number of cells in your original sample, you would have to then multiply by your dilution factor.

How would you calculate the number of bacteria/1.0 ml of *diluted* sample using the grid shown in Figure 14.2? Assume that there is an average count of 11 cells per square. Did you get 44,000 bacteria/1.0 ml of *diluted* sample?

B. DIRECT MICROSCOPE COUNT

In the direct microscope count, a known volume of the suspension is smeared on a slide over a known surface area, usually 1 cm^2. After the suspension dries, the cells can be stained and counted. This method is commonly used to count bacteria in milk and is often referred to as the *Breed count*. In this procedure, 0.01 ml of the milk is spread over an area of exactly one cm^2. The number of bacteria in several randomly chosen fields of view are counted and the mean number of bacteria per field calculated. If you know the area of the field of view (from Exercise 3), you can calculate the number of bacteria in the square centimeter over which the 0.01 ml of suspension was spread by dividing 1.0 cm^2 by the area of the field of view. This will give you the number of fields/cm^2. Multiply the number of fields/cm^2 by the average number of bacteria/field to obtain the number of bacteria in 0.01 ml of the suspension you are counting. Multiply by 100 to obtain the number of bacteria per milliliter.

LABORATORY OBJECTIVES

The microscope can be used to count the number of cells in a suspension. To do this, you must:

- Understand the use of counting chambers and the relationship between the surface area of a grid and the volume of the space above it.

- Understand how the number of cells in a volume of fluid can be determined by the Breed-count procedure.

- Understand that the direct counting method will often, but not always, provide data that cannot be obtained by the plate count method.

- Appreciate that most experimental methods often have limitations that preclude their use in all instances.

MATERIALS NEEDED FOR THIS LABORATORY

A. COUNTING CHAMBERS

1. Cell-counting chambers (either Neubauer or Petroff-Hauser chambers). Be sure to note which is being used, and therefore the correct measurements of the grid, the number of subdivisions, and the depth of the counting chamber.

2. Coverslip to be used with the counting chamber.

3. Pipettes calibrated to accurately deliver 0.01 ml.

4. Suspension of Baker's Yeast.

5. A 24-hour heat inactivated broth culture of *nonmotile* bacteria such as *Bacillus megaterium*.

6. Test tube containing 9.0 ml water.

7. Methylene blue stain.

8. Crystal violet stain.

B. DIRECT MICROSCOPIC COUNT

1. Breed-counting slides. If Breed slides are not available, you can use a glass slide as indicated in the procedures.

2. Pipettes calibrated to accurately deliver 0.01 ml.

3. A 24-hour heat inactivated broth culture of *nonmotile* bacteria such as *Bacillus megaterium*.

4. Milk which is unpasteurized or pasteurized milk which has been kept overnight at room temperature.

5. Test tubes containing 9.0 ml of water

6. Xylol

7. 95% alcohol

8. Methylene blue stain.

LABORATORY PROCEDURES

1. Obtain a counting chamber and the coverslip that is to be used with it. (Why is this coverslip so much thicker than other coverslips we have used?) Using distilled water, carefully clean the counting surface and the coverslip.

YEAST SUSPENSION

2. Prepare a dilution of yeast suspension by pipetting 1.0 ml of a well mixed yeast suspension into a tube containing 9.0 ml of water. What is the dilution of your yeast suspension? Add several drops of methylene blue stain to the suspension.

3. Draw a small amount of the diluted yeast suspension into a pipette. Allow some of the suspension to form a bead on the pipette tip. Touch the tip to the edge of the counting cham

ber. (Some counting chambers have a triangular groove cut into their surface for the pipette tip.)

4. The droplet will be drawn into the chamber by capillary action and should just fill it. If there is too much fluid in the chamber, it will spill over into the grooves. Any excess fluid should be removed with absorbent paper.

5. Allow the counting chamber to stand for a few moments.

6. Gently place the counting chamber on the microscope stage, with the counting chamber centered over the light source. Locate the grid with the low power of the microscope and position it so that you can count the number of cells in one of the large squares.

7. Change magnification as needed. Count the number of cells in at least ten squares and calculate the mean number of cells per square.

8. Complete the necessary calculations to determine the number of yeast cells per ml of original suspension.

9. Record your results on the Laboratory Report Form.

BACTERIAL SUSPENSION

10. Mix one loopful of crystal violet with 1.0 ml of a 24-hour broth culture.

11. Follow steps 3 – 8 as indicated for the yeast suspension.

12. Record these results on the Laboratory Report Form.

B. DIRECT MICROSCOPIC COUNT

1. Obtain a Breed-counting slide. It will contain from one to five marked staining areas. Each area is 1 cm². If a Breed slide is not available, you can approximate one by marking off 1 cm squares with a wax pencil. Use a centimeter ruler to draw a square centimeter on a piece of paper. Gently warm a clean slide over the Bunsen burner and then position it over the square. Quickly (before the slide cools) outline the square centimeter on the slide with a wax pencil.

   You may avoid using the wax pencil by using a loop to spread the drop of fluid over a cm² area, using the drawing on the paper as a template.

BACTERIAL SUSPENSION

2. Use a pipette to accurately transfer 0.01 ml of bacterial suspension to one of the squares on the slide. If the slide is clean, the drop will spread evenly over the glass and fill the marked-off area. If necessary, use a loop to spread the fluid.

3. Make both a 1:10 and 1:100 dilution of the bacterial suspension and repeat step 2, applying the suspensions to different staining squares.

4. Allow the smears to air dry. Heat fix the smears over a boiling-water bath for a few minutes.

5. Stain the smears with methylene blue for 30 seconds.

6. Rinse briefly with distilled water.

7. Allow the slides to air dry before attempting to count the bacteria.

8. Using the oil-immersion objective, count the number of bacteria in each of at least ten fields of view. Calculate the mean number of bacteria per field.

9. Calculate the number of bacteria per milliliter of original suspension.

10. Record your results on the Laboratory Report Form.

MILK SAMPLE

11. Use a pipette to accurately transfer 0.01 ml of milk to one of the squares on the slide. If the slide is clean, the drop will spread evenly over the glass and fill the marked-off area. If necessary, use a loop to spread the fluid.

12. Allow the smear to air dry. Heat fix the smear over a boiling-water bath for a few minutes.

13. Flood the slide with xylol and then rinse gently with 95% alcohol. The xylol removes the fat in milk and the alcohol washes any residual xylol from the slide.

14. Stain the smear with methyiene blue for 30 seconds.

15. Decolorize gently with alcohol until the smear appears light blue. Rinse briefly with distilled water.

16. Allow the slide to air dry before attempting to count the bacteria.

17. Using the oil-immersion objective, count the number of bacteria in each of at least ten fields of view. Calculate the mean number of bacteria per field.

18. Calculate the number of bacteria per milliliter of milk.

19. Record your results on the Laboratory Report Form.

LABORATORY REPORT FORM

EXERCISE 18
INTRODUCTION TO MICROSCOPY

What was the purpose of this exercise?

A. COUNTING CHAMBERS

Record the following information about the counting chamber you used.

Type of counting chamber: _____
(Neubauer or Petroff-Hauser)
Length of one side of the center grid: _____
(usually bound by double lines)
Area of center grid: _____

Depth of counting chamber: _____

Volume of space over center grid: _____

Length of side of medium squares: _____
(usually bound by heavy lines)
Number of medium squares per mm^2:   _____

Length of side of smallest squares: _____
(usually bound by thin lines)
Number of smallest squares per mm^2:   _____

YEAST SUSPENSION:

Number of cells per square: _ _ _ _ _ _ _ _ _ _

Mean number of cells per square: _ _ _ _ _ _ _ _ _ _

Calculate the number of cells per ml of your original yeast suspension. Be sure to show all of your work.

BACTERIAL SUSPENSION:

Number of cells per square: __ __ __ __ __ __ __ __ __ __

Mean number of cells per square: __ __ __ __ __ __ __ __ __ __

Calculate the number of cells per ml of your original yeast suspension. Be sure to show all of your work.

B. DIRECT MICROSCOPIC COUNT

Record the following information for your microscope:

Area of oil-immersion field: _____
(Refer to results for Ex. 3)

BACTERIAL SUSPENSION:

Dilution counted: _____

Number of cells per field: __ __ __ __ __ __ __ __ __ __

Mean number of cells per field: __ __ __ __ __ __ __ __ __ __

Calculate the number of cells per ml of your original bacterial suspension. Be sure to show all of your work.

If you performed both the counting chamber and direct count methods on the same bacterial suspension, how did your results compare?

MILK SAMPLE:

Number of cells per field: __ __ __ __ __ __ __ __ __ __

Mean number of cells per field: __ __ __ __ __ __ __ __ __ __

Calculate the number of cells per ml of your original milk sample. Be sure to show all of your work.

QUESTIONS:

1. List three circumstances where a direct microscope count would be preferred over a plate count.

2. If you made an error in counting your cells resulting in a mean count of one additional cell, how much error (in other words, how many more cells) would your results include? Use the figures for your bacterial suspension results for comparison.

3. The Neubauer counting chamber is used to determine the exact number of bacteria present in a urine sample. The sample has been diluted 1:200 and an average count of 19 bacteria/field has been obtained. What would be the number of bacteria per ml of urine?

4. What would be the significance of the above urine count if the urine sample had been allowed to sit at room temperature for several hours following collection before the count were performed?

MICROBIAL POPULATION COUNTS:
VIABLE CELL COUNTS

The size of a bacterial population can be estimated by measuring some biological activity of the population or by plating the cells on a suitable medium and allowing them to grow into visible colonies. The term *viable cell count* refers to the counting of cells by plating them on a nutrient medium and counting the colonies that develop. Often the results are referred to as *plate counts* or *colony counts* and the results reported as *colony-forming units*.

Colony counts have some distinct advantages when compared to direct microscope counts. The colony count is significantly more sensitive than the direct count. Some microbiologists claim that there must be more than 50,000 bacteria in each milliliter before you can reliably and accurately count them on a microscope slide. On the other hand, plate-counting procedures, particularly when membrane-filtration techniques are used, can detect a very few colony forming units in relatively large volumes of liquid. When manually counting colonies, it is important that there be at least 30 colonies on the plate before the count is considered to be a representative sample. Why?

BACKGROUND

SOLUTIONS, SUSPENSIONS, AND DILUTIONS

A word or two on terminology. A solution results when one chemical (the *solute*) is dissolved in another, usually liquid, chemical (the *solvent*). A good example would be a solution of sodium chloride (salt) in water. The salt, being soluble in water, is the solute, while the water is the solvent. The salt is truly soluble in water. A *suspension* results when one of the components is not soluble in the liquid. If you added sand or dirt to water, you would have a suspension of dirt or sand in water. Usually, but not always, solutions are clear while suspensions are turbid.

It is often necessary to further dilute suspensions and solutions. The dilutions are expressed as relative volumes, such as 1 to 10 or 1 to 100 (written as 1:10 or 1: 100). Dilutions are made by using pure diluent to change the volume of the sample according to the relative volumes desired. For example, if you wanted to make a 1:10 dilution, you would change the volume from 1.0 ml to 10.0 ml by adding 1.0 ml of the suspension or solution to 9.0 ml of diluent. Using the salt solution as an example, if you transferred 1 ml of a 1% solution to 9 ml of water, you would have made a 1:10 dilution of the original solution. Also, the concentration of the salt would have been reduced by a factor of 10.

Most bacterial and viral samples can be treated as suspensions. If the sample is liquid, we determine the number of bacteria or viruses suspended in the liquid, diluting it as needed. If it is solid, you must first make a suspension (typically, 10% in water) by suspending the sample in a liquid. When these samples are counted, the concentration of the initial suspension must be calculated into the dilution factor. For example, a 1.0% suspension has already been diluted 1:100.

SERIAL DILUTIONS

One of the most important quantitative techniques routinely used in clinical microbiology laboratories is the serial dilution procedure. When a sample has been diluted out, each of the dilutions can be tested to determine which of them give test positive results. Alternatively, the number of bacteria or viruses in each of the diluted samples can be counted after they grow on a suitable medium.

When the number of bacteria or viruses in a sample must be counted, the report usually indicates the number of bacteria or viruses in each milliliter of the original sample. The report might read "15,000 bacteria/ml" or "1 50,000 phage/ml." If very large numbers of bacteria are encountered, the logarithmic number is often used instead of the arithmetic number. This, of course, does not change the results at all, but it does make things somewhat more convenient. The actual number of bacteria or viruses is determined by multiplying the number of colonies (or plaques) observed on the plate times the dilution(assuming 1.0 ml aliquots were plated out).

number of bacteria/ml =
number of colonies X dilution

A serial dilution is typically a stepwise dilution sequence made by transferring aliquots from one tube of diluent to another. The aliquots are usually the same volume throughout the sequence. If a 1:10 dilution sequence is used, the process is referred to as a tenfold serial dilution; if a 1:2 dilution sequence is used, it is a twofold serial dilution. The number (ten, two, etc.) is the dilution accomplished at each step, a tenfold serial dilution being a series of 1:10 dilutions. The following table has some examples:

Serial	Dilution	Volumes Used
		Aliquot transferred into
Tenfold	1:10	1.0 ml into 9.0 ml
Fivefold	1:5	1.0 ml into 4.0 ml
Twofold	1:2	1.0 ml into 1.0 ml

Since a serial dilution is a series of identical, stepwise dilutions, the final dilution, or the dilution of any tube in the sequence, can be easily determined. The actual dilution sequences are arithmetic or logarithmic progressions, the interval or "step" being determined by the dilution used at each step. The highest dilution is always in the last tube, while the lowest dilution is always in the first tube. Using zero (0) to indicate the undiluted sample is sometimes helpful in keeping track of the dilution steps. When bacterial or viral suspensions with very large numbers or organisms are being diluted, the undiluted sample is frequently not counted. When antibody titre in serum is to be determined, or when the expected numbers are very low, the undiluted sample is often counted or assayed.

POUR-PLATE COUNTING OF BACTERIA

If you added 1.0 ml of a bacterial suspension to melted agar and then poured the agar into a petri dish, the bacteria would be immobilized in the agar when it solidified. After a suitable incubation time, the colonies would become visible and you can count them (see Exercise 8). **Figure 19.1** shows a typical ten-fold serial dilution/pour-plate counting protocol.

Although a single colony may contain more than a million bacterial cells, it can be assumed that all of those cells arose from a single cell. The number of colonies is, therefore, an accurate representation of the number of viable cells (able to grow on the medium used and under the incubation conditions used) present in the sample that was mixed with the agar.

If you did one of these counts on each tube in a serial dilution, you would have completed a plate count. Your results might look like those in Table 19.1.

When counting colonies, only those plates that have between 30 and 300 colonies are counted. In the example shown in the table, the only "countable" plate is plate #5. The reason for counting plates that have between 30 and 300 colonies is that the most representative samples are obtained when the number of colonies falls within that range. Less than 30 colonies is not considered to be a representative sample (too few organisms). More than 300 colonies is similarly considered nonrepresentative because the overcrowding of the colonies will result in poor growth of the bacteria, and some colonies may not have grown to visible size. Furthermore, technician error increases as the colonies become smaller, more numerous, and more difficult to see.

The results from the plate count are given as the number of colonyforming units per ml of original culture (CFU/ml). To calculate this value, it is necessary to multiply the actual colony count by the dilution factor. For example, for plate #5 above, 38

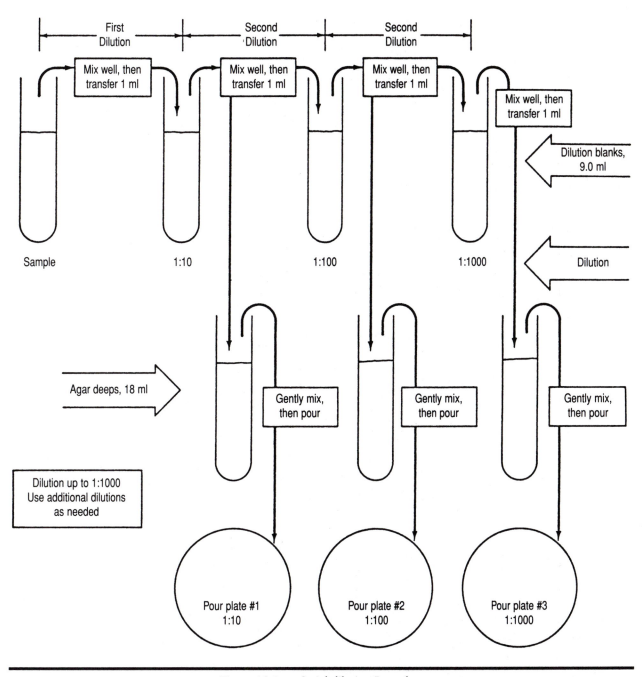

Figure 19.1   Serial-dilution Procedure

colonies were counted in a 1:100,000 (10^{-5}) dilution. When the actual count is multiplied by the dilution factor you get a value of 3,800,000 CFU/ml. This can also be represented using the log-10 value which would be 3.8 X 10^6.

Table 19.1   Results of a Bacterial Plate Count

PLATE NUMBER	DILUTION	LOG 10 DILUTION	COLONY COUNT
1	1:10	-1	TNTC*
2	1:100	-2	TNTC
3	1:1000	-3	TNTC
4	1:10,000	-4	321
5	1:100,000	-5	38
6	1:1 million	-6	5
7	1:10 million	-7	NONE

*Indicates colonies were TooNumerous to Count

MEMBRANE-FILTER COUNTING

As you have seen, serial dilutions are useful when the number of bacteria in a sample is very large. Membrane-filter counting techniques make it possible to detect and accurately count small numbers of bacteria in relatively large samples. Also, since the bacteria are not exposed to the heat of melted agar, this procedure is useful for counting heat-sensitive organisms. Membrane-filter counting has proven particularly useful in the bacteriological testing of water (where large volumes can be easily tested for low numbers of coliforms) and in quality-control laboratories (where large volumes can be tested to be sure that they are sterile).

When a known volume of the sample is passed through a bacteriological-grade filter, any bacteria in the fluid will be retained on the filter. These bacteria will remain viable for a short period of time and will develop into visible colonies if the filter is placed on a pad saturated with nutrient medium. The volume of sample tested is limited only by the capacity of the filtering apparatus.

This technique is very adaptable to specialized counting procedures and, depending upon the type of medium used, can differentially count bacterial populations. For example, membrane-filter counting is routinely used to detect coliforms in water samples containing other bacteria.

CONSIDERATIONS USED IN PLATE COUNTING

Certain assumptions and considerations have to be taken into account in all plate-counting procedures. These include:

1. Each colony is assumed to be the progeny of a single cell. If we could be assured that each colony did arise from a single cell, then the assumption that the number of colonies equals the number of cells in the suspension could be accepted. In fact, while it is true that some types of bacteria do readily dissociate into single cells, many do not.

The variability in the tendency of bacterial cells to remain clumped will introduce some error to your counts unless you are careful to treat each sample the same. For example, you should mix the dilution blanks in a consistent manner (e.g., 50 shakes, one minute on a vortex mixer at a given speed, etc.).

Because of this tendency for organisms to remain clumped, many microbiologists prefer to use the term *colony-forming units per milliliter* (CFU/ml) instead of bacteria per milliliter. They point out that it is the ability of the cells to grow and to exercise their usual biological activity that is of concern. It really doesn't matter whether the colony (or infectious unit) arose from a single cell or not; what does matter is that it will grow into a colony or may cause an infection.

2. Except for relatively rare and unusual circumstances, plate counts of naturally occurring populations actually select for relatively small segments of these populations. The growth medium that is used in the petri plates as well as the conditions under which the plates are incubated are themselves selective.

Pure cultures of laboratory strains of bacteria, under some circumstances, represent an example where all cells might be expected to grow when transferred to a petri plate containing an appropriate nutrient medium.

Most natural populations of bacteria, such as would be found in the intestinal tract, in water samples, or in soil samples, contain bacteria that are not able to grow in the environmental conditions used for plate counting. Often special media must be used, or other environmental conditions must be manipulated. These include light, pH, anaerobiosis, temperature, special sources of oxidizable carbohydrate, and so on.

3. Frequently, however, the selectivity of the medium or the growth environment can be used to encourage the growth of specific kinds of bacteria. For example, we might want to know how many anaerobic bacteria were in the population. In soil microbiology it is important to know how large the cellulose-decomposing segment of the population is. In water, we need to know the number of coliforms before we can determine whether the water is safe to drink.

In clinical microbiology, the presence of more than 100, 000 bacteria in urine is considered significant, but in most other cases, where a normal flora is usually present, we are only interested in certain types of bacteria (beta-hemolytic, lactose nonfermenters, etc.).

4. The application of membranefiltration techniques to plate-counting procedures makes it possible to greatly increase the sensitivity of such counts. For example, it is possible to pass large volumes of fluid through a filter, trapping the bacteria on the filter. If the filter is then placed on a filter-paper pad that has been saturated with medium and then incubated, the bacterial cells will grow into distinct colonies. In theory, any bacteria (even one) in the entire volume of fluid that passed through the filter will appear as a colony after suitable incubation. The technique is not only very sensitive, but the results can be obtained sooner. These techniques have found wide application in the quality testing of water, foods, and medicines.

5. Finally, it is often necessary to perform both a direct microscope count and a plate count on a sample. What would you conclude from a milk sample that had a plate count of 1,000 CFU/ml and a direct count of 25,000 bacteria/ml?

SOURCES OF ERROR IN SERIAL DILUTIONS

Two common sources of error introduced in the serial-dilution procedure are pipetting errors and errors introduced by failing to mix the diluted samples sufficiently. The pipetting errors tend to resolve themselves with practice. As you gain experience, the

reproducibility of your results will improve. If you remember a few rules about pipetting, your errors will be markedly reduced.

1. Always determine the type of pipette you are using and then use it correctly. A measuring pipette is designed to accurately deliver the volume indicated between two marks on the pipette. The graduations on a measuring pipette do not extend to the tip. A serological pipette is designed to deliver the stated volume by allowing the fluid to flow out of the pipette under the force of gravity. The graduations on this type of pipette always extend to the tip.

2. Always hold the pipette in the vertical position. This way, the meniscus can always be read more accurately.

3. Always use the bottom or top of the meniscus (do not try to estimate the midpoint) to measure the volume. Because of the surface-tension properties of aqueous solutions, the meniscus formed in a pipette is usually convex, and the bottom of the meniscus should be used as your measuring point.

Mixing errors can be avoided by taking care to always mix the samples carefully. You should mix all the tubes the same way—for example, by vortex mixing for one minute, or by shaking twenty times. It really does not matter how you mix the sample, as long as you are consistent and careful to ensure complete mixing.

1. Always mix the dilutions carefully.

2. If a vortex mixer is available, vortex each tube for the same amount of time.

3. If you use bottles or screw-cap tubes, shake each bottle or tube the same number of times.

4. The key to accurate dilutions is consistency; mix and pipette each dilution the same way.

The serial-dilution protocol will use 1.0 ml aliquots for plating. It is sometimes convenient to plate out 0.1 ml and make the appropriate mathematical changes in dilution calculation. Using the smaller volume allows you to stretch out your dilution sequence, but it also makes pipetting errors proportionately larger.

SOME REMINDERS

Plate counts usually require that the sample be diluted before plating. Study the diagram in Figure 19.1. The reason for the dilution of the sample is to ensure that you will have plates with between 30 and 300 colonies on them. As a starting point, a 24-hour culture of *E. coli* will contain a cell population of between 1 and 100 *million* cells per milliliter.

SOME APPLICATIONS

While it is often important to know the size of a bacterial population, it is sometimes even more important (and interesting) to learn how that population changes. For example, what would be the effect of changing the nutrients in a medium? How would you go about testing which medium produced better growth—nutrient broth or brain-heart-infusion broth?

What results would you expect to obtain if you did a plate count on a broth culture every two or three hours for about 24 hours? Would you be able to demonstrate the typical growth curve? Why not try it?

How would you evaluate the quality of water or milk? Most foods must meet standards that include limits on the numbers of bacteria per gram or milliliter. Often they will specify limits that include total bacteria as well as total coliforms. How would you modify this exercise to detect coliforms? (Look at Exercise 35 for possible suggestions.)

LABORATORY OBJECTIVES

Viable cell counts determine the number of living cells in a population. Although this information may be most important, there are some limitations on the procedure. In this exercise you should:

- Understand the relationship between the serial dilution and the plating of the sample.

- Be able to list several of the limitations to the plate-count procedure.

- Understand the limitations and advantages of membrane-filter counts.

- Be able to discuss how plate counting and direct counting can often provide different, yet complementary, information about the population being studied.

MATERIALS NEEDED FOR THIS LABORATORY

A. POUR-PLATE COUNTS

1. 9.0 ml water or saline (0.9% NaCl) dilution tubes—7 tubes will be needed if a tenfold serial dilution to 1:10 million is to be performed.

2. Nutrient agar deeps—3 – 7 tubes. If you are going to bracket the expected number of colonies and produce at least one dilution with between 30 and 300 colonies, you will need 3 tubes. If you do not know about how many colonies to expect, you should plate out more dilutions.

3. Sterile 1.0 ml pipettes. You will need at least one pipette for each dilution step.

4. Sterile petri plates.

5. Sample to be plated. Any available source such as water, soil, milk, or a food product may be used. Eighteen to 24 hour broth cultures of *Staphylococcus epidermidis* or *Escherichia coli* with populations of about 1 million bacteria/ml may also be used.

B. MEMBRANE-FILTER COUNTS

1. Membrane-filter apparatus. This should include a receiving funnel or vessel, filter holder, and suction funnel. Complete assemblies are available commercially. These usually include all components, including the filter, in a single assembly.

2. Vacuum source or pump.

3. Sterile membrane filters, 0.45 μm pore size, to fit membrane-filter apparatus.

4. Sterile filter-paper absorbent pads, same diameter as filters

5. Sterile petri plate.

6. Sterile, fresh m-Endo MF broth.

7. Sterile 5.0 ml pipettes.

8. 20 ml of sterile water for rinsing filters.

9. Water samples, including at least one that is known to contain coliforms. Students might use varied samples, with adjacent groups testing coliform-free and contaminated samples.

LABORATORY PROCEDURE

A. POUR-PLATE PROCEDURE

1. Obtain and label all the dilution blanks, nutrient agar deeps, pipettes and petri plates you will need to complete your dilutions and platings. Remember to label the petri plates so that your labels will not obstruct your view through the bottom of the plate.

2. When melted, maintain the agar deeps in a water bath (at about 50°C) to prevent the medium from solidifying before you use it.

3. Arrange your material so that everything is within reach.

4. Decide which of the dilutions will be plated out. If the culture or sample you are working with has about 10,000,000 cells/ml, you should at least plate out the 1: 100,000 through the 10,000,000 dilutions. If necessary, ask your instructor for help.

5. Following the diagram in Figure 19.1, complete your serial dilutions and plating simultaneously. Use the same pipette to transfer 1 ml from one dilution blank to the melted agar and to the next dilution blank.

6. Allow the poured medium to solidify before moving the plates.

7. Incubate the plates for 24 hours at 37°C.

8. Count the number of colonies on each plate. If the medium is transparent, you should count the colonies through the bottom of the plates. It is sometimes helpful to mark each colony with a felt-tip pen as you count it.

9. Use any plates that have between 30 and 300 colonies to calculate the number of colony-forming units (CFU) per milliliter in the original sample. Don't forget to take into account your dilution steps. You do not need to allow for the dilution of the sample by the medium. Why not?

10. Record all results and calculations on the Laboratory Report Form.

B. MEMBRANE-FILTER TECHNIQUES

1. Using Figure 19. 2, assemble a membrane-filtration apparatus. Clamp the assembly together, being careful that there are no leaks around the filter. If you are using a prepared apparatus, follow the instructions that came with it. Remember to carefully observe aseptic techniques, especially when handling the filter.

2. Using aseptic techniques, place an absorbent pad into a petri plate. Add the required amount of nutrient medium—about 2 ml. Set the plate aside while you complete the filtration steps.

3. Pour the water sample into the receiving funnel and apply vacuum to the suction flask. Be sure to record the volume of water filtered. After the sample passes through the filter, rinse the sides of the receiving funnel with about 20 ml of sterile water or saline.

4. Turn the vacuum source off and gently remove the vacuum hose. Allow a minute or two for the air pressure in the suction flask to equilibrate. If you leave the vacuum on for too long, you may kill the cells by excessive drying.

5. Carefully remove the clamp and receiving funnel. If necessary, use sterilized forceps to separate the filter from the bottom of the receiving funnel.

6. Using sterile forceps, transfer the filter into the culture plate, centering it over the saturated nutrient pad. Be sure that there is no air trapped between the filter and the nutrient pad. The medium will be drawn up into the filter by capillary action.

7. Close the petri plate. If you are using specially designed filter plates the cover will fit tightly. Incubate for 22 to 24 hours. Do not invert the petri plate.

8. Count the colonies that appear on the filter. If you used m-Endo MF broth, conforms will produce a distinctive golden to green metallic sheen

9. Record your results on the Laboratory Report Form.

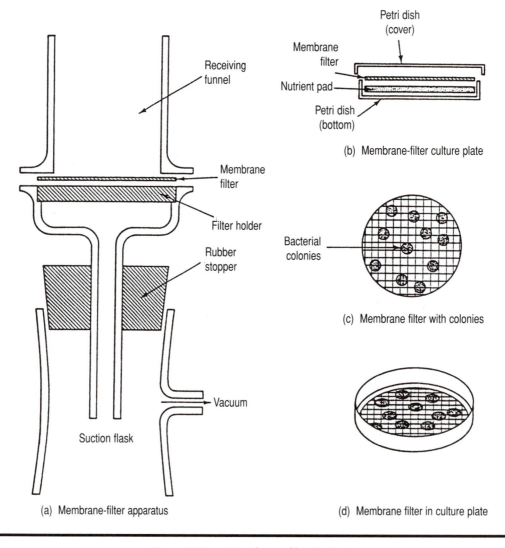

(a) Membrane-filter apparatus

(b) Membrane-filter culture plate

(c) Membrane filter with colonies

(d) Membrane filter in culture plate

Figure 19.2   Membrane-filter Technique

LABORATORY REPORT FORM

EXERCISE 19
MICROBIAL POPULATION COUNTS

What was the purpose of this exercise?

A. PLATE COUNT RESULTS

Report your plate-count results on the following table. Give the actual number of colonies counted at each dilution and calculate the number of CFU/ml in the original sample. Finally, convert the counts to their logarithmic numbers.

PLATE COUNT RESULTS

OBSERVED DATA		CALCULATED RESULTS	
DILUTION	COLONIES COUNTED	ACTUAL COUNT (CFU/ml)	LOG 10 OF COUNT (CFU/ml)
MEAN FOR ALL COUNTS:			

B. MEMBRANE-FILTRATION COUNT

Report the following data for the sample you counted by membrane filtration.

Volume of water filtered: _____

Coliform colonies per milliliter: _____

Coliform colonies per sample: _____

QUESTIONS:

1. List two circumstances when the standard plate-count procedure would be the method of choice.

2. List three advantages of the membrane-filter technique.

3. Why are counts above 300 and below 30 statistically unreliable?

4. What factors would contribute to these techniques being selective? How could their selective nature be minimized?

5. The number of bacteria in a milk sample has been determined by preparing a ten-fold serial dilution series. Five dilution tubes were utilized, and the final three dilutions were plated. The first plate was too numerous to count (TNTC), the second one had 198 colonies and the third plate had 23 colonies. Determine the number of bacteria in the original sample (CFU/ml).

SPECTROPHOTOMETRIC METHODS

As you are now well aware, bacterial suspensions vary in turbidity; when you inoculate a tube of broth it becomes increasingly turbid as the culture matures. You may already suspect that the more bacteria present in the suspension, the more turbid the suspension will be.

In this exercise you will attempt to determine, by constructing an absorbance-cell number standard curve, if there is a direct relationship between the number of bacteria per milliliter and the turbidity of the suspension. You will then use the curve to estimate the amount of growth that occurs in a culture over the time of your laboratory period.

BACKGROUND

Bacteria, when suspended in a clear fluid, both scatter and absorb light. The amount of light absorbed or scattered is proportional to the density of the bacterial suspension. In other words, the amount of light absorbed or scattered can be graphically related to the number of bacteria in the suspension. Some modern spectrophotometers do not directly measure the light that is absorbed or scattered, but rather measure the amount of light that actually passes through (is not absorbed or scattered by) the sample.

The relative opacity of the sample is reported as either percent transmission (%T) or absorbance. Percent transmission indicates the percent of light that passes through the sample—that is, the light that is neither absorbed nor scattered. Absorbance, on the other hand, indicates how much light *is* absorbed or scattered.

SPECTROPHOTOMETERS

A spectrophotometer is a device that can measure the intensity of light. Study **Figure 20.1**. The *light source* produces light over a wide range of wavelengths in either the visible or ultraviolet range of the spectrum. Some spectrophotometers have more than one light source that can be switched into or out of the light path, as needed. A lens system is needed to focus the light. The *filter* or *defraction grating* selects the proper wavelength of light while eliminating others. The light entering the sample will be, for all practical purposes, monochromatic. This is important because different chemicals absorb different wavelengths of light, and if the wrong wavelength is used, no absorbance will be observed.

The *photoelectric cell* is a special tube that converts light energy into electrical energy. When such a tube is correctly connected to a galvanometer, the amount of electrical energy can be measured and can be shown to be directly proportional to the amount of light striking the photoelectric cell.

To use the spectrophotometer you should adjust the galvanometer to show a transmittance of 100% or an absorbance of 0.00 when a tube containing sterile, clear broth is in the sample holder. Then, as shown in **Figure 20.2**, when a tube containing turbid broth (such as a culture of bacteria) is inserted into the sample holder, the galvanometer will register a different reading because some of the light will have been scattered or absorbed by the bacterial cells. Typically, the meter will show a lower percent transmission or a higher absorbance.

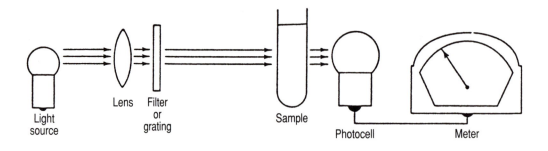

Figure 20.1   Diagram of a single-cell spectrophotometer.

STANDARD CURVES AND HOW TO USE THEM

If you were to plot absorbance or percent transmission against the number of bacteria per milliliter of sample, you would, over at least part of the curve, obtain a line that appears to be a straight line. Furthermore, if you kept the general conditions constant (the species of bacteria, composition of medium, etc.) The curve would be reproducible. This type of curve is frequently referred to as a *standard curve*.

Once the graph was drawn, it could be used to determine the number of bacteria per milliliter by simply reading the percent transmission or absorbance of the suspension. It is only necessary to measure the absorbance of the suspension, then determine its coordinates from the graph, to find the number of bacteria per milliliter. Many samples can be read rapidly, saving time and material.

Standard curves, such as the one described here, can be used to estimate the density of various bacterial suspensions. For example, they are frequently used to determine the weight of cells at the beginning and end of growth experiments, to complete bacterial growth curves, and for bioassay procedures where the amount of bacterial growth during a specified incubation period must be determined .

In constructing the graph for your standard curve, you must take into account the logarithmic nature of absorbance. A graph that plots absorbance (on the y-axis) against the number of cells (on the x-axis) should be constructed on arithmetically ruled paper, whereas one that plots percent transmittance against cell number should be drawn on semi-logarithmic paper, with the percent transmittance shown on the logarithmic scale.

LABORATORY OBJECTIVES

There is a "direct" relationship between the number of cells in a suspension and the absorbance of light by that suspension. This relationship can be used to estimate the number of cells in the suspension by using a spectrophotometer to measure absorbance accurately and comparing the reading with a previously constructed standard curve. To be able to do this you should:

- Understand the relationship between the density of the cells in the suspension and the absorbance and/or scattering of light by that suspension.

- Understand the indirect relationship between percent transmission (%T) and absorbance.

- Understand the use of standard curves and how once the relationships in the objectives above are known, the data can be used to determine the density of cells by simple photometric measurements (instead of plate counts).

MATERIALS NEEDED FOR THIS LABORATORY

1. Bausch & Lomb Spectronic 20 spectrophotometer.

2. Spectrophotometer tubes or cuvettes.

3. Kimwipes or other lint-free tissues .

A. CALIBRATION OF THE SPECTROPHOTOMETER

1. One tube of sterile nutrient broth.

B. CONSTRUCTION OF A STANDARD CURVE

1. A 24-hour culture of *E. coli*. You will need a volume of about 10 ml. The culture must be at a maximum turbidity. Use nutrient broth supplemented with 0.1% yeast extract and 1.0% glucose.

2. Five tubes, each containing 4.0 ml of sterile nutrient broth.

3. Sterile pipettes, 10.0 ml and 1.0 ml

4. 9.0 ml dilution blanks to dilute a culture to 1: 1 0 million

5. Four nutrient agar deeps

6. Four sterile petri plates

7. Container of disinfectant

8. Ice bath

C. MEASUREMENT OF BACTERIAL GROWTH BY ABSORBANCE

1. A 24-hour culture of *E. coli*. You will need a volume of about 10 ml. The culture must be at a maximum turbidity. Use nutrient broth supplemented with 0.1% yeast extract and 1.0% glucose.

2. One culture tube containing 8.0 ml of sterile nutrient broth supplemented with 0.1% yeast extract and 1.0% glucose.

3. Sterile pipettes, 1.0 ml.

4. 37°C water bath

IV. LABORATORY PROCEDURES

Before you begin the experimental procedure, you must calibrate the spectrophotometer and prepare the cuvettes.

Turn on the spectrophotometer and let it warm up for at least 30 minutes.

CARE OF CUVETTES

1. Cuvettes should be washed before use by rinsing with distilled water. If spots remain, wash with a gentle detergent solution and rinse several times with distilled water.

2. Hold the cuvette only by the upper third of the tube. If you must mark the cuvettes, do so only at the top and only with a wax pencil.

3. Rinse the cuvette with distilled water between each reading. Allow the tube to drain by hold-

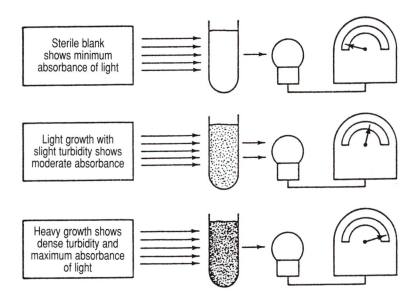

Figure 20.2   Effect of turbidity on absorbance.

ing a Kimwipe against the open end while the cuvette is inverted .

4.   Wipe the outside of the cuvette with Kimwipes, never with a paper towel.

5.   Always insert the cuvette into the sample holder with the index marks on both the cuvette and holder aligned.

A.   CALIBRATION OF THE SPECTROPHOTO-METER

1.   Set the wavelength knob (see **Figure 20.3**) to a wavelength between 550 and 650 nm. Using the zero adjust knob, adjust the meter to read exactly zero. **NOTE:** The same wavelength is to be used for the entire exercise.

2.   Pour 4 ml of sterile nutrient broth into a cuvette. Wipe the bottom half with a Kimwipe then insert it into the sample holder. The index on the tube should be aligned with the index mark on the tube holder.

3.   Close the cover of the sample holder and adjust the meter to read 100% transmittance by rotating the light-control knob. Remove the sample of nutrient broth.

4.   If the meter does not return to zero transmittance when the cuvette is removed from the holder, you must repeat steps 2, 3, and 4.

B.   CONSTRUCTION OF A STANDARD CURVE

NOTE: The two parts of this exercise may be completed simultaneously. If you are doing both Parts A and B in the same laboratory period, you should set up Part B first and complete Part A while Part B is incubating. The spectrophotometric readings for Part B should be taken while you are constructing the standard curve for Part A.

1.   Obtain five culture tubes containing 4.0 ml of sterile nutrient broth. Label them "1:2" through "1:32" respectively. Also obtain one empty sterile tube; label it "Undiluted"
     NOTE: Bacteria will be able to grow quite well in the nutrient broth even at room temperature. If the absorbance readings are not taken at about the same time as the plate counts are completed, significant error can be introduced. To minimize this error, ice-cold broth should be used to make the dilutions and the tubes should

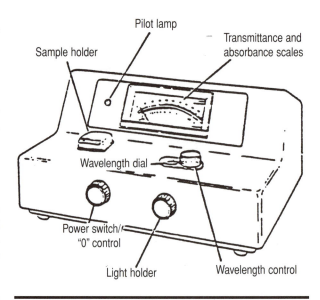

Figure 20.3   Bausch & Lomb Spec20

be held in an ice bath until the experiment is completed .

2.   Study the protocol shown in **Figure 20.4**

3.   Transfer 4.0 ml of your sample culture to the tube labeled "undiluted" and to the tube labeled "1 :2."

4.   Mix the first dilution (1:2) by gently swirling and then transfer 4.0 ml from it to the next dilution tube (the one labeled "1:4"). Mix well.

5.   Continue the serial-dilution sequence until all dilutions have been completed (the last one will be the 1:32 dilution). Discard 4.0 ml from he last tube into a container of disinfectant. **NOTE:** You should determine the absorbance of the suspensions in their reverse order. That is, read the most dilute suspension (1:32) first, working your way up to the least dilute (undiluted) suspension. This will minimize error due to carryover of suspension from one sample to another.

6.   Confirm that the spectrophotometer has been calibrated. (Refer to the Calibration of the Spectrophotometer section.) Pour the contents of the last tube into a cuvette and place the cuvette into the sample holder. Close the cover and record the percent transmittance.

7. After you determine the percent transmission of the sample, pour the broth back into the tube. (Save each sample until the experiment is complete, just in case you have to do it over.) Rinse the cuvette with distilled water, discarding the rinse water into a container of disinfectant.

8. Repeat steps 6 and 7 until you have measured the percent transmittance of all the tubes, including the undiluted suspension. Blank the spectrophotometer between each reading by ascertaining that a reading of 100%T is obtained with sterile nutrient broth.

9. Record all your readings on your Laboratory Report Form. Calculate the absorbance for each reading.

10. Assemble and label the materials needed to complete a plate count. You will need to dilute your sample to 1:10 million and plate out the four highest dilutions. Review the procedure in Exercise 19.

11. Complete a plate count on the undiluted suspension. Use 1 ml of the first tube ("Undiluted") as your sample.

C. MEASUREMENT OF BACTERIAL GROWTH BY ABSORBANCE

1. Preheat the *Escherichia coli* culture and sterile supplemented nutrient broth to 37°C prior to starting this exercise.

2. Obtain a tube containing about 8.0 ml of sterile supplemented nutrient broth. If possible, use a spectrophotometer cuvette or a new culture tube that shows little if any absorbance with sterile nutrient broth between 550 and 650 nm.

3. Add 1.0 ml of the 24-hour culture of *E. coli* to the sterile broth. Determine the absorbance of this suspension.

4. If this suspension shows a percent transmittance greater than 80%, add an additional amount of the broth culture until the transmit-

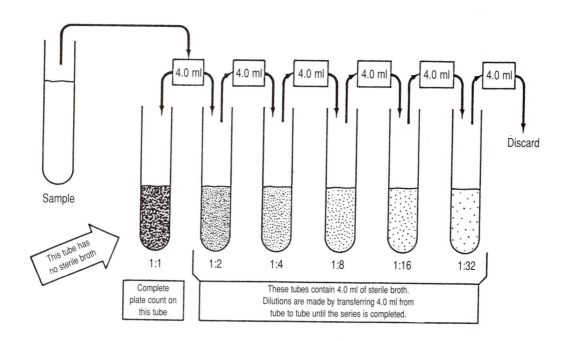

Figure 20.4   Twofold serial dilution for standard curve (absorbance vs. bacterial count).

tance is less than 80%. Record the reading in the report form. **NOTE:** Step 3 is very important. If the culture is not dense enough, you will be unable to detect changes in the transmittance during the three-hour incubation time. If necessary, you should incubate the culture until its percent transmittance is less than 80%.

5. Incubate the suspension in a water bath at 37°C for the remainder of the laboratory period (or at least three hours), taking the percent transmittance every thirty minutes. Record these readings on your Laboratory Report Form .

LABORATORY REPORT FORM

EXERCISE 20
SPECTROPHOTOMETRIC METHODS

What was the purpose of this exercise?

B. CONSTRUCTION OF A STANDARD CURVE

1. Complete the following table:

TURBIDITY-ABSORBANCE STANDARD CURVE			
DILUTION	CFU/ml*	%T	ABSORBANCE**
Undiluted			
1:2			
1:4			
1:8			
1:16			
1:32			

*To be determined from your plate counts (See #2).
**To be calculated using the formula   Absorbance = 2 - (log of %T) or
                                             Absorbance = log (100%/T)

2. Count the colonies on all countable plates and calculate the number of colony-forming units per milliliter (CFU/ml) for each dilution, including the undiluted sample.

3. Plot the absorbance/turbidity by graphing with the optical density on the y-axis and dilution on the x-axis. Use the graphs on the following page.

Change in absorbance over time

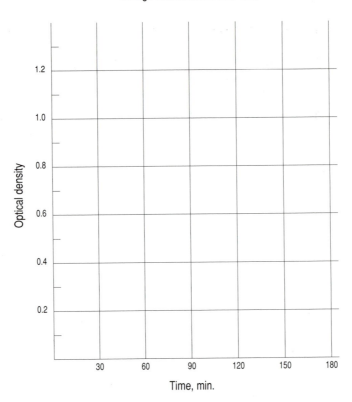

Time, min.

Standard curve: absorbance/CFU/ml

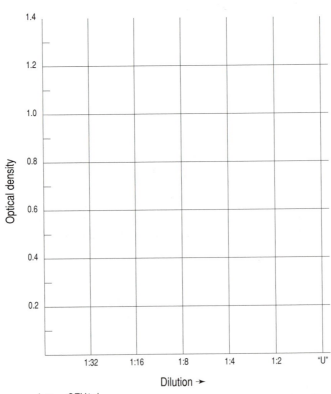

Dilution ➝

Log_{10} CFU/ml — — — — — — — — — — —

4. Complete the following table. You will need to use the data developed for the standard curve.

BACTERIAL GROWTH (Measured Spectrophotometrically)			
TIME	%T	ABSORBANCE	CFU/ml
Zero			
+ 30 min.			
+ 60 min.			
+ 90 min.			
+ 120 min.			
+ 150 min.			
+ 180 min.			

5. Plot the absorbance of the suspension against time.

QUESTIONS:

1. Why did you use nutrient broth and not distilled water as the reference blank?

2. Of what value are the curves you developed? How might a microbiologist use them?

3. You tool %T readings for your culture over a period of three hours and plotted those values. What would you predict would happen, both to the %T values and the graph, if you were to continue to read these values over the next several days.

PHYSICAL CONTROL METHODS

In Exercise 10, we examined the physical growth requirements of several organisms. Just as we use our knowledge of an organism's optimal physical growth requirements to guarantee its successful cultivation, we can use this information to inhibit the organism's growth.

BACKGROUND

Each organism has an optimum temperature, osmotic pressure, oxygen level, and pH. In addition, all living cells require the maintenance of adequate water concentrations to survive. Often, these organisms are able to survive for at least limited periods of time in the absence of optimal conditions. Contributing to an organism's survival rate under less than optimal conditions would be the presence of an endospore, the growth phase of the organism, and the nutrients present.

Microbial control methods include *sterilization, disinfection,* and *sanitization.* Sterilization is defined as the destruction or removal of all microbial organisms and their products in or on an object. Disinfection, on the other hand, is the destruction, inactivation or removal of microorganisms likely to cause undesirable effects such as disease or spoilage. (Does a disinfectant sterilize?) To sanitize means to simply reduce the number of organisms present with no guarantee of their destruction.

We often cannot depend only upon sanitization, but need to be able to guarantee the destruction of certain microorganisms. This might be in the sterilization of bacteriologic media or surgical instru-

ments, the disinfection of the laboratory bench or an operating room, the antisepsis of a patient's skin prior to surgery or following a cut, or the preservation of our food. It requires that we have some knowledge of the organisms that we are most likely dealing with and the conditions under which they can be destroyed. In this exercise we will examine the role of several physical conditions in microbial control.

TEMPERATURE

We saw in Exercise 10 that each organism has an optimum temperature as well as a range in which it survives. Each organism also has a *thermal death time* and *thermal death point* which can be determined. The thermal death time is the shortest period of time required to kill a suspension of a given microorganism at a given temperature and under specified conditions. It is important to note that this must be determined for each species. The thermal death point, on the other hand, is the lowest temperature at which a given microorganism is killed in a given time (usually 10 minutes). Knowledge of these values are important in determining the conditions that will be used for the control of microorganisms by heat.

Heat has a number of applications in microbial control. This involves both moist heat (autoclaving, pasteurization, or boiling-water disinfection) and dry heat (hot-air-oven and incineration) methods. No matter which method is used, its ability to control microbial growth is dependent upon its interference with normal enzymatic activity. When enzymes are heated too greatly, they can no longer maintain their proper shape and become inactive, a process known as denaturation of protein. Without proper enzyme functioning, cells cannot function and death occurs. The difference in the amount of heat needed and its overall effectiveness will vary dependent upon both the ability of the heat to penetrate the cell and the presence of resistant structures such as endospores.

It should also be noted that enzyme activity slows down as the temperature of an organism decreases. This, too, can serve to diminish the catalytic capabilities of the enzyme and inhibit the growth of the cell. The lowering of temperature, however, is not a guarantee of cellular death. (Can meat taken out of the freezer still spoil even though it does not come in contact with new contaminants?)

In this exercise you will determine the thermal death time for an organism and also examine the effect that freezing temperature has on microbial growth.

OSMOTIC PRESSURE

We also saw in Exercise 10 that each organism has a range of osmotic pressure in which it could survive or actively grow. Hypotonic environments are rarely lethal to microorganisms due to the presence of a rigid cell wall which resists the in flow of too much water and the rupturing of the cell. (What does the term hypotonic mean? If you are not sure, go back to Exercise 10 and reread the background information.) Hypertonic solutions are far more effective in microbial control. In fact, hypertonic solutions were the first major methods of food preservation. Perhaps the greatest hypertonic environment we can present an organism is that which occurs in the total absence of water or when drying occurs. Why is this considered hypertonic? What would be the effect on the bacterial cell? How do we use various hypertonic solutions today to preserve food? Remember, this can involve the increased concentration of either salt or sugar or the decreased concentration of water.

OBJECTIVES

In this exercise, we will be examining the effect that elevated temperature, lowered temperature, salt concentration and drying have on microbial growth. When performing this exercise, you should:

- Propose, perform, evaluate, and appraise the results of an experiment that establishes thermal death time.

- Examine the relationship between the physical growth requirements of an organism and the physical methods of microbial control.

MATERIALS NEEDED FOR THIS LABORATORY

A. THERMAL DEATH TIME

1. Two TSA plates.

2. One sterile test tube.

3. 24-hour broth culture of *Bacillus subtilis*

4. 100°C water bath.

5. Test tube rack

B. THE EFFECT OF FREEZING

1. 24-hour broth culture of:
    Escherichia coli
    Bacillus subtilis
    Staphylococcus aureus

2. Three sterile test tubes

3. Two TSA plates

4. Sterile 5.0 ml pipettes and pipette bulb

C. THE EFFECT OF OSMOTIC PRESSURE

1. Broth cultures of:
    Escherichia coli
    Bacillus subtilis
    Staphylococcus aureus
    Halobacterium salinarium

2. TSA tall (15 ml) containing 15% NaCl

3. TSA tall (15 ml) containing 0.5% NaCl

4. Sterile petri plate

5. 50°C water bath

D. THE EFFECT OF DRYING

1. 24-hour broth cultures of:
    Escherichia coli
    Bacillus subtilis
    Staphylococcus aureus

2. Three sterile screw-capped test tubes.

3. Three sterile swabs

4. Disinfectant

5. Sterile nutrient broth

6. Sterile 5.0 ml pipettes

PROCEDURES

A. THERMAL DEATH TIME

Remember, thermal death time is the amount of time required to kill bacteria and their spores at a given temperature. You are going to design an experiment to determine the thermal death time of *Bacillus subtilis* using the scientific method. This requires you to:

    a. state the problem;
    b. formulate a testable hypothesis;
    c. make a prediction;
    d. design and perform an experiment, being sure to incorporate adequate controls;
    e. form a conclusion based on experimental results;
    f. re-evaluate experimental design; and
    g. if necessary, formulate another testable hypothesis and test it.

The following limitations are placed on your experimental design:

    a. The test organism will be *Bacillus subtilis*.
    b. The temperature used for your experiment will be 100°C.
    c. The equipment available will include:
        two TSA plates
        one sterile test tube
        100°C water bath
        test tube rack

1. Determine your experimental design. Remember, TSA plates can be divided into varied number of sectors. Be sure that your experiment is designed to answer the initial question! Record your experimental design on the Laboratory Report Form.

2. Perform the experiment as designed. Be sure to include appropriate controls.

3. Incubate the plates at 37°C until the next laboratory period.

4. Examine your plates. Did you determine the thermal death time for *Bacillus subtilis?* What changes would you make if you were to repeat the experiment?

B. THE EFFECT OF FREEZING

1. Obtain a TSA plate. Divide the bottom into three sectors labeling each with the name of one of your test organisms.

2. Inoculate each sector with one of the broth cultures. (What is the purpose of this plate?)

3. Obtain 3 sterile test tubes. Label each with the name of one of the organisms.

4. Carefully pipette 3.0 ml of the first broth culture into its labeled sterile tube. Repeat with new sterile pipettes for each organism.

5. Place the tubes into the freezer until the next laboratory period.

6. Remove the cultures from the freezer and allow them to thaw. They may be placed in tepid water to thaw.

7. Observe the plate from Day One. Record the presence or absence of growth for each organism on your Laboratory Report Form.

8. Obtain a TSA plate. Divide it into three sectors and label one sector for each organism.

9. Inoculate each sector with the appropriate thawed culture.

10. Incubate at 37°C until the next laboratory period.

11. Following incubation, observe the plate for growth. Record your results on the Laboratory Report Form.

C. THE EFFECT OF OSMOTIC PRESSURE

1. Place a tube of melted TSA tall containing 15% NaCl and a tube of melted TSA tall containing 0.5% NaCl in a 50°C water bath.

2. Put a pencil on your laboratory bench. Place a sterile petri plate with one edge resting on the pencil. (See **Figure 21.1**)

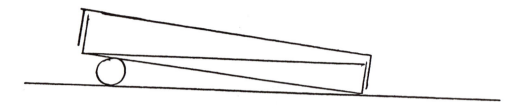

Figure 21.1   Preparation of Gradient Plate

3. Carefully add the TSA containing 15% NaCl so that it does not quite cover the bottom of the plate. Cover the plate with its lid and leave it setting on the pencil until the agar sets (approximately 15 minutes).

4. When the bottom wedge of agar has solidified, label the side of the plate with the thickest portion of agar "15%." Place the plate flat on the laboratory bench. Carefully pour the TSA tall containing 0.5% NaCl over the surface of the agar wedge until it just flows over the thickest portion of the wedge. Replace the cover on the plate and let it sit until it is completely solidified.

5. After the plate is solidified, draw four evenly spaced parallel lines from the 15% NaCl side to the 0.5% NaCl side of the plate. Label each line with the name of one of the test organisms. (See **Figure 21.2**)

6. Using your broth cultures, inoculate each along its marked line, starting at the low concentration end of the line.

7. Incubate the plate at 37°C until the next laboratory period.

8. Following incubation, record the growth patterns observed on your Laboratory Report Form.

D. THE EFFECT OF DRYING

1. Obtain 3 sterile screw-capped test tubes, three swabs and the broth cultures. Label one tube for each organism being tested.

2. Dip a sterile swab into a broth culture. Be sure to ring the inside of the tube with the swab so the swab is not too wet.

3. Swab the inside of the labeled screw-capped tube with the culture. Be careful not to leave a large drop in the tube.

4. Repeat steps 2-3 with the other organisms.

5. Place all tubes in the 37°C incubator until the next class period.

6. Pipette 3.0 ml of sterile nutrient broth into each of the screw capped tubes.

7. Re-incubate until the next laboratory period.

8. Observe each tube for the presence of growth.

9. Record your results on the Laboratory Report Form.

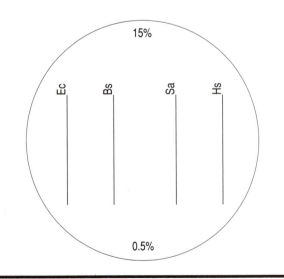

Figure 21.2   Gradient Plate

LABORATORY REPORT FORM

EXERCISE 21
PHYSICAL CONTROL METHODS

What was the purpose of this exercise?

A. THERMAL DEATH TIME

1. What is your hypothesis about the thermal death time of *Bacillus subtilis?*

2. What is your experimental design? Be sure to include adequate controls!

3. Report your results for your experiment.

4. Did you determine the thermal death time for *Bacillus subtilis?*

   If so, what is it? _____

   If not, what would you vary if you were to repeat the experiment?

B. EFFECT OF FREEZING

1. Complete the following table indicating the presence (+) or absence (−) of growth.

   Freezing Time: _____

ORGANISM	CONTROL	FOLLOWING FREEZING
Escherichia coli		
Bacillus subtilis		
Stephylococcus aureus		

2. What effect did freezing have on the survival of the test organisms?

C. OSMOTIC PRESSURE

1. Record your results on the following diagram:

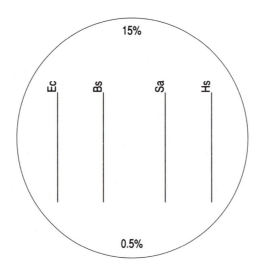

2. Approximately what concentration of sodium chloride, if any, was inhibitory to the test organisms?

Escherichia coli _____

Bacillus subtilis _____

Staphylococcus aureus _____

Halobacterium aelinarium _____

D. EFFECT OF DRYING

1. Complete the following table indicating the presence (+) or absence (–) of growth.

ORGANISM	CONTROL	FOLLOWING DRYING
Escherichia coli		
Bacillus subtilis		
Stephylococcus aureus		

2. What effect did drying have on the survival of the test organisms?

QUESTIONS:

1. How would you expect the thermal death time for *Escherichia coli* to differ from that for *Bacillus subtilis*. Why?

2. Why is *Clostridium botulinum* a major concern in the canning of food?

3. Culture strips of *Bacillus subtilis* or *Bacillus sterothermophilus* are often used as quality control systems for autoclaving. Why are these organisms used?

4. Based on the results from this exercise, propose methods by which you could:

   Sterilize meat _____

   Preserve meat _____

   Disinfect green beans _____

ULTRAVIOLET LIGHT AS AN ANTIMICROBIAL AGENT

CAUTION: Ultraviolet light can damage the retina of the human eye. *Under no circumstances should you look into the lamp used in this experiment. Avoid having reflected light shine into your eyes.* Avoid direct exposure of the skin with the ultraviolet light.

Ultraviolet light is often used to disinfect or sterilize surfaces of objects that cannot be conveniently sterilized by other methods. It is also used to sterilize air entering and leaving certain types of "clean rooms." Its effectiveness as a bactericidal agent is based on the light-absorbing properties of nucleic acids.

BACKGROUND

THEORETICAL BASIS FOR BACTERICIDAL EFFECTS OF ULTRAVIOLET LIGHT

Sunlight is composed of a continuous spectrum of electromagnetic radiation with varying wavelengths (See **Figure 22.1**). Visible light consists of those wavelengths between 400 and 900 nm (a nanometer or nm = 10^{-9} m). Those wavelengths above the visible spectrum are the infrared rays, microwaves and radio waves. Those below 400 nm fall into the ultraviolet rays, X-rays, and gamma rays. In this exercise, we are most concerned about the wavelengths in the ultraviolet range (100 – 400 nm). Ultraviolet rays below 200 nm are readily absorbed by air. The ultraviolet rays that exhibit the greatest affect on cells, specifically the DNA, are those between 200 and 290

nm, or the UVC wavelengths. The wavelengths between 250 and 260 nm have been found to be the most injurious to cells.

The ultraviolet rays stimulate the formation of bonds between the carbons of adjacent pyrimidines (See **Figure 22.2**). The resultant *pyrimidine dimers* interfere with the proper pairing of complementary bases, thereby interfering with proper DNA replication by the incorporation of improper nucleotides. This results in mutation or, in severe cases, death of the cell. It should be noted that not all ultraviolet-induced mutations of DNA are lethal or detrimental. Mutations can also have a beneficial or neutral effect on the viability of the affected cell.

Cells have two mechanisms that enable them to reverse the effect of ultraviolet light on the DNA— *photoreactivation* and *dark repair* (See **Figure 22.3**). Photoreactivation, or light repair, occurs in the presence of visible light. Light activates an enzyme which breaks the bond in the pyrimidine dimer, returning DNA to its original state.

Dark repair, which does not occur in all bacteria, involves the utilization of several enzymes. Endonucleases "cut outs" the dimer, DNA polymerase synthesizes new DNA to replace this section, and ligase attaches the new segment to the original strand.

USE OF ULTRAVIOLET LIGHT AS A BACTERICIDAL AGENT

There are a number of limits to the application of ultraviolet light as an antimicrobial agent. Exposure of items does not always result in sterilization.

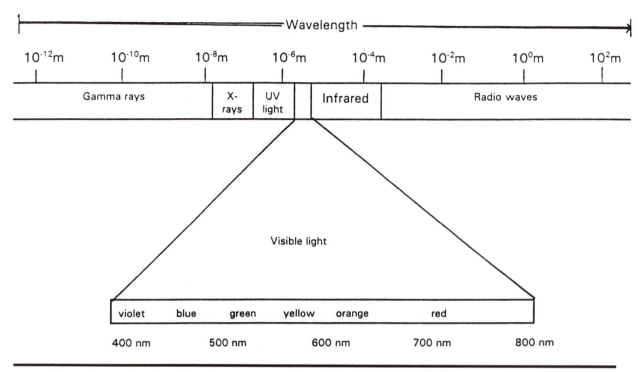

Figure 22.1   Light Spectrum

Ultraviolet rays have very poor penetrating abilities and their effect is readily blocked by glass, water, and the presence of other organic material. Pigmented and endospore-forming organisms are more resistant to its effects than are actively growing non-pigmented organisms. [What hypothesis can you form as an explanation for this?] The effectiveness of ultraviolet light is influenced by the distance between the source of the light and the cells. The closer the light to the cells, the more intense its effect. In addition, there is the problem of both photoreactivation and dark repair which can reverse the effects of the ultraviolet light.

Another problem that must be addressed before ultraviolet light can be used is its safe application. The effect of ultraviolet light on DNA is not restricted to bacterial cells and bacterial DNA. The same effects occur in any cell undergoing direct exposure to ultraviolet light [Why is our overexposure to sunlight discouraged?]. It is recommended that individuals not be present when ultraviolet lights are being used or that the lights are adequately shielded to protect individuals in the area.

Ultraviolet light has been used for the destruction of airborne organisms in operating rooms, pharmaceutical companies, research laboratories, and food processing areas—but primarily in the absence of people. It has also been used for the sterilization of heat labile solutions, and disinfection of wastewater.

EXPERIMENTAL DESIGN

This experiment is designed to demonstrate the germicidal properties of ultraviolet light and the possible effects that components of the medium may have in the protection of organisms from the ultraviolet radiation. To accomplish this, you will suspend bacteria in two different solutions—saline and nutrient broth. They will be exposed to ultraviolet rays and, at certain time intervals, a loopful of the irradiated suspensions will be transferred to fresh nutrient agar. Any bacteria that have survived exposure will produce visible growth after incubation.

Since saline does not contain any ultraviolet-absorbing material, the bacteria should be killed rapidly. The nutrient broth contains ultraviolet absorbing components and should show some protective effects.

LABORATORY OBJECTIVES

In this exercise you will determine the time needed to kill suspensions of bacteria exposed to ultraviolet radiation. To complete this exercise, you will need to

- Understand the mechanism by which ultraviolet radiation accomplishes it bactericidal effects.

- Explain the environmental limits to the use of ultraviolet radiation.

- Understand the reasons why certain components of the medium may protect suspended bacteria from ultraviolet-induced lethal effects.

- Discuss commercial and medical applications of ultraviolet-light disinfection.

- Identify two repair mechanisms that enable cells to survive exposure to ultraviolet radiation.

- Demonstrate the effect of endospores and pigment on the effectiveness of ultraviolet radiation.

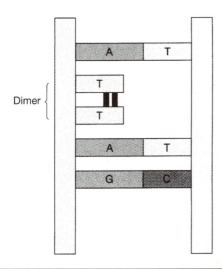

Figure 22.2   Formation of Pyrimidine Dimer

MATERIALS NEEDED FOR THIS LABORATORY

1. Four TSA plates per group.

2. Broth cultures of the following:
   Escherichia coli (24-hour)
   Serratia marcescens (24-hour)
   Bacillus subtilis (72-hour)
   Each group will be assigned one of the organisms to use for the laboratory procedure.

3. One tube each of sterile saline.

4. One tube of double-strength nutrient broth.

5. Two empty, sterile petri dishes. If available, use the small (60 mm diameter) size.

6. Sterile pipettes.

7. Ultraviolet light with a wavelength between 250 and 260 nm located 6 – 9" above the surface.

LABORATORY PROCEDURE

1. Obtain four TSA plates. Divide each plate into four quadrants by marking on the bottom with a wax pencil or marker.

2. Label two of the plates "Saline" and the other two "Nutrient broth." Label the quadrants on the first saline plate "0," "30 sec.," "1 min." and "2 min." The second saline plate should be labeled "3 min.," "5 min.," "10 min.," and "15 min." Label the nutrient broth plates in the same manner.

3. Each group will be assigned one of the organisms to use.

4. Transfer 1 ml of the assigned broth culture to the tube of saline and 1 ml to the tube of nutrient broth. Mix well.

5. Pour about 3 ml of the mixtures into separate, empty petri plates.

6. Transfer one loopful of each bacterial suspension in the petri plates to the quadrant of the plates labeled "0." These will serve as your unexposed controls.

A WORD OF CAUTION: Ultraviolet light can damage the retina of the human eye. *Under no circumstances should you look into the lamp used in this*

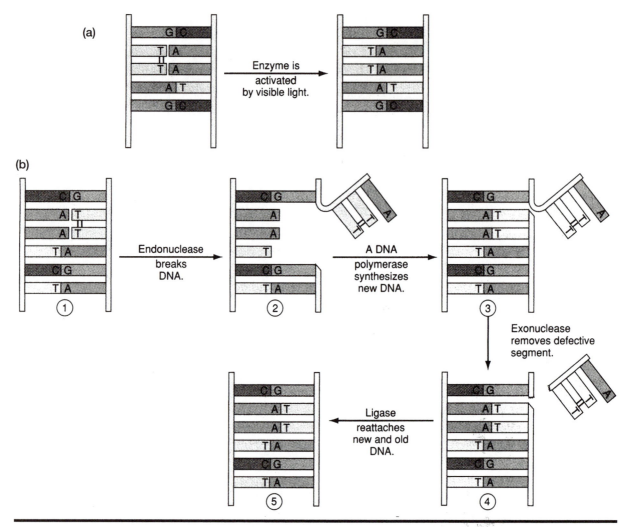

Figure 22.3   Repair of Pyrimidine Dimers: (a) Photoreactivation; (b) Dark Repair.

experiment. Avoid having reflected light shine into your eyes. Avoid direct exposure of the skin with the ultraviolet light.

7. Place the open petri plates (they must be open because the ultraviolet light may be absorbed by the cover) under the ultraviolet light source. Be sure that the entire plate is exposed to the light. Begin timing immediately after putting the plates under the light source.

8. At the designated times, transfer one loopful of each suspension to the appropriate quadrant of the petri dish containing the nutrient agar. The saline suspension will be transferred to the plates labeled "Saline" and the nutrient broth suspension will be transferred to the plates labeled "Nutrient Broth."

Exposure Times:
    30 sec.
    1 min.
    2 min.
    3 min.
    5 min.
    10 min.
    15 min.

9. Incubate plates as indicated:
    Escherichia coli—37°C
    Bacillus subtilis—37°C
    Serratia marcescens—room temperature

10. Observe the plates for growth. With the *Serratia marcescens*, also note the presence of nonpigmented colonies.

11. Record your results on the Laboratory Report Form.

LABORATORY REPORT FORM

EXERCISE 22
ULTRAVIOLET LIGHT AS A BACTERICIDAL AGENT

What was the purpose of this exercise?

1. What was the distance between the ultraviolet light source and the petri plates?

2. Complete the following table. Be sure to get the data for the organisms you did not test from others in your class. Record the amount of growth as 4+ (heaviest growth), 3+, 2+, 1+ (slight growth), or 0 (no growth).

TIME	Escherichia coli		Serratia marcescens		Bacillus subtilis	
	SALINE	BROTH	SALINE	BROTH	SALINE	BROTH
0						
30 sec.						
1 min.						
2 min.						
3 min.						
5 min.						
10 min.						
15 min.						

3. Did you observe any variation in pigmentation in *Serratia marcescens* following exposure to ultraviolet light? If so, what variation did you see and in which plate(s)?

4. How might you explain any pigment variation observed with *Serratia marcescens*?

QUESTIONS:

1. Why must the petri plates be left open during the exposure to ultraviolet light?

2. How is DNA damaged by ultraviolet light?

3. Explain photoreactivation.

4. How could you modify this experiment to determine the mutagenic activity of ultraviolet light?

5. What other forms of radiation are lethal to microbes? Are they used as antimicrobial agents? Why or why not?

6. Why is the distance between the light source and the surface of the fluid an important variable in this exercise? What would happen if you decreased the light intensity?

7. *Staphylococcus aureus,* which is frequently found on the skin, is pigmented, while *Escherichia coli,* isolated fromt he intestines, is not. Based on this experiment, what possible explanation can you give?

8. Why can't you use ultraviolet lights to sterilize bacteriological media?

ASSAY OF ANTIMICROBIAL AGENTS:
DISK-DIFFUSION METHODS

How effective are the antiseptics and disinfectants you use? Does silver nitrate really work when used to rinse the eyes of newborns? Can you effectively sterilize the counter or tabletop? Can the antiseptic used to cleanse the skin also be used to sterilize instruments? These and many other questions must be answered if effective sterilization and disinfection methods are to be used.

There are methods that can be used to evaluate the effectiveness of antimicrobial agents. However, before we look at the specific methods of evaluation, lets review some of the terminology commonly used to discuss these agents and how they inhibit or destroy microorganisms.

BACKGROUND

Antimicrobial agents include all those agents, physical or chemical, that destroy or inhibit microorganisms. We examined the effectiveness of several physical control agents in Exercises 21 and 22. In this exercise, as well as in Exercises 24 and 25, we will concentrate on the effect of chemical control agents on microbial growth.

The term germicide is often used to refer to a chemical capable of destroying microorganisms. Increased specificity of action can be shown by the use of more precise terms, such as bactericide (an agent that kills bacterial cells, but not necessarily spores), sporicide (an agent that kills bacterial endospores or fungal spores), virucide (an agent which kills viruses), or fungicide (an agent that kills

fungi). The distinction made by these terms is important in determining their possible application. For example, what type of agent would you want to use to kill the mildew in your shower?

Other chemicals do not kill microorganisms, but simply inhibit their growth. These agents are referred to as being microbistatic. Again, it is possible to determine greater specificity of action by the utilization of the term *bacteriostatic* to describe those agents that specifically inhibit the growth of bacterial cells.

Chemical control agents are also classified by their ability to be used on living tissue in contrast to only being applicable to inanimate surfaces. Those chemicals that are applied to living tissue for the prevention of infection (antisepsis) are commonly referred to as being antiseptics. They must be relatively nontoxic, allowing their safe application to skin or mucous membranes and are usually, but not always, bacteriostatic. Disinfectants, on the other hand, are used on inanimate surfaces. These may be the same agents as antiseptics, but are usually used in a higher concentration. Would household bleach be considered an antiseptic or a disinfectant? How about hydrogen peroxide?

There is a third group of chemical control agents which will be considered in Exercise 25—the *chemotherapeutic agents*. These chemicals are used specifically to treat disease. They consist of antibiotics, or those chemicals derived from living agents, and synthetic drugs. We will discuss these further in Exercise 25.

FACTORS AFFECTING ANTIMICROBIAL ACTIVITY

Several factors must be taken into consideration when choosing antimicrobial agents as they can affect their effectiveness. What is the degree of microbial contamination? Are endospores present? How long will the organisms be exposed to the antimicrobial agent? What is the temperature of application? Does the microbial environment contain extraneous organic material? What is the pH of the microbial environment?

CLASSIFICATION OF ANTIMICROBIAL AGENTS

Antimicrobial agents are classified by their chemical composition and method of action. The major groups, their action and effectiveness are summarized in Table 23.1.

Usually when you receive an injection or have blood drawn, the skin is cleansed with an alcohol wipe. Because of the rapid evaporation rate, this has somewhat limited antimicrobial effectiveness, simply killing vegetative cells that are on the surface of the skin. Endospores, resistant bacteria, and organisms within the skin pores are not affected. Alcohol serves as a lipid solvent, dissolving cell membranes. When it is coupled with water it also denatures protein. For this reason, ethyl or isopropyl alcohol are most commonly used in a 70 – 80% concentration.

Another widely used skin antiseptic as well as disinfecting agent are the halogens—primarily iodine and chlorine. Tincture of iodine was one of the first antiseptic agents used. It has been replaced in clinical applications by iodophors which combine the iodine with organic surfacants. This provides a longer period of antisepsis. Betadine and Isodine are commonly used iodophors for surgical scrubs and preparation of skin prior to incisions. Chlorine is used as hypochlorous acid in bleach, water treatments and various commercial sanitizing agents. One major limitation to its application, however, is that it may be inactivated by the presence of organic material.

The heavy metals—mercury, copper, silver, zinc, and selenium—work by oligodynamic ("little power") action, combining with protein molecules and denaturing them. Frequently used forms of mercury include merthiolate and mercurochrome for skin antisepsis, and trimerosal for skin antisepsis, instrument disinfection and vaccine preservation. Copper is most commonly used as copper sulfate for

the control of algal growth in bodies of water. Silver nitrate has been used to prevent the growth of gonococcal organisms in the eyes of newborns and both silver nitrate and silver sulfadiazine are used for severe burns. Both zinc and selenium are included as antifungal agents in skin ointments and shampoos.

Lister first used phenol as an antimicrobial agent. Since then, phenol has been replaced by a number phenol derivatives or phenolics due to the toxicity of phenol. These agents denature protein and disrupt the membranes of cells. Their effectiveness is not affected by the presence of organic material. Many of the phenolics, such as hexachlorophene and chlorhexidine gluconate, are combinations of phenolics and halogens, providing increased effectiveness.

Many of the commonly used laboratory disinfectants are quaternary ammonium compounds. Included are Zephiran and Roccal. Their use is now somewhat limited, however, as they have been shown to not only fail to inhibit *Pseudomonas* organisms, but to support their growth.

METHODS OF EVALUATION OF ANTIMICROBIAL AGENTS

There are three major ways of evaluating the effectiveness of antimicrobial agents. Each of the techniques has certain advantages and certain disadvantages. All are useful under certain circumstances. Each of these methods is summarized below.

Disk-Diffusion Technique

The disk-diffusion technique is commonly used to test the overall effectiveness of antiseptics, disinfectants and chemotherapeutic agents. Its application for chemotherapeutic agents will be used in Exercise 25 (the Kirby-Bauer procedure). The test is based upon the diffusion of molecules (the agent being tested) from the point of high concentration (the disk) to low concentration (the media). The effectiveness of the agent diminishes as the concentration decreases, resulting in varied zones of inhibition (See Figure 23.1). The rate of molecular diffusion is dependent upon several factors including the molecular weight and shape of the substance being tested, its concentration, the media being used, the presence of organic material, and the temperature of incubation. For this reason one cannot make the assumption that the larger the zone of inhibition, the more effective an agent is, but only whether or not the chemical agent will inhibit the given organism. In addition, it is not

Table 23.1   Antimicrobial Agents and Their Applications

Antimicrobial	Action	Effectiveness	Examples	Applications
Acids	Metabolic inhibition, lower pH, denature protein	Widely used as fungicide	Sorbic acid Benzoic acid Calcium propionate	Food preservation, antifungal agents
Alcohols	Denature proteins, dissolve lipids, dehydrate molecules	Kill vegetable cells but not endospores	Ethanol and isopropanol	Disinfect instruments and clean skin
Alkylating agents	Raise pH, denature protein	Kill vegetable cells and endospores	Formaldehyde Glutaraldehyde	Embalming and vaccines Antiseptic (Cidex)
Dyes	May interfere with replication or block cell wall synthesis	Bacteriostatic in high concentrations, at lower concentrations only inhibits gram-positive organisms	Acridine Crystal violet	Clean wounds Treat some protozoan and fungal infections, Selective agent in bacteriological media
Gaseous Chemosterilants	Denature protein	Sterilizing agent, especially for materials which would be harmed by heat	Ethylene glycol Propolene glycol Ethylene oxide	Aerosols Sterilize inanimate objects that would be harmed by high temperatures
Halogens	Inactivates protein, oxidizes cell components	Kills vegetative cells and some endospores	Iodine Chlorine	Antiseptic; surgical preparation (Betadine) Disinfection for water, dairies, restaurants
Heavy metals	Denature protein	Disinfectant and antiseptic applications dependent upon concentration	Silver nitrate Mercury compounds Copper	Prevent gonococcal infections; Disinfect skin and inanimate objects; Inhibit algal growth
Oxidizing agents	Oxidation, disrupt disulfide bonds	Especially effective against oxygen-sensitive anaerobes	Hydrogen peroxide Potassium permanganate Ozone	Clean puncture wounds Disinfect instruments Water disinfection
Phenolics	Denature protein, disrupt cell membranes	Kill vegetative cells but not endospores, many phenolics effective in presence of organics	Cresols Hexachlorophene Chlorhexidine gluconate	Preservatives Antiseptic Surgical scrub
Quaternary ammonium compounds	Denature protein, disrupt cell membranes	Kill most vegetative cells	Zephiran, Roccal, other cationic detergents	Sanitization in restaurants, laboratories, industry
Soaps and detergents	Lower surface tension	Alone, primarily sanitizing agent, may be coupled with other antimicrobial agent(s)		Hand washing, laundering, sanitizing kitchen and dairy equipment

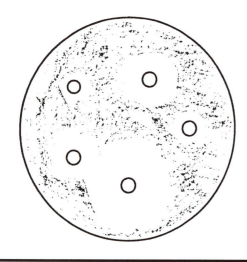

Figure 23.1   Zones of Inhibition

possible to determine whether the agent is bactericidal (killing the microorganism) or simply bacteriostatic (inhibiting the organisms' growth). The disk-diffusion test is the easiest procedure to use when one is simply trying to determine the overall effectiveness of a given agent against a given organism.

Use-Dilution Technique

In the use-dilution procedure, bacteria are removed from the inhibitory agent, enabling the determination of its bactericidal or bacteriostatic action. Varied dilutions of the antimicrobial agent can be used to determine the **minimal inhibitory concentration (MIC)** or **minimal bactericidal concentration (MBC)** of the agent. Because this technique is more complex than the disk-diffusion method, it is not routinely performed. It would be used for the establishment of effectiveness for a newly formulated antimicrobial, or clinically when patients do not respond to what is normally considered adequate therapy or who have a relapse while undergoing adequate therapy. We will utilize a modified use-dilution test in Exercise 24.

Phenol Coefficient Test

The phenol coefficient test determines the ratio of the concentration of antimicrobial agent being tested to the concentration of phenol that will kill the bacteria being tested in ten minutes, but not in five minutes. This provides a quantitative comparison of the tested agent and phenol, however the value of this information is limited. It must be remembered that the phenol coefficient value determined is significant

only for the organism tested under the test conditions used and should not be interpreted more broadly. It also does not take into account the presence of any extraneous organic material. We will not be performing the phenol coefficient test, however the following information should help you understand its performance and use.

Disinfectant "X" and phenol are tested to determine their effectiveness against *Staphylococcus aureus*. The following results are determined:

DISINFECTANT "X"

Concentration	5 minutes	10 minutes
1:100	-	-
1:150	-	-
1:200	+	-
1:250	+	-
1:300	+	+

PHENOL

Concentration	5 minutes	10 minutes
1:20	-	-
1:30	-	-
1:40	+	-
1:50	+	-
1:60	+	+

To determine the phenol coefficient, you would determine the greatest dilution of Disinfectant "X" that kills in 10 minutes but not 5 minutes (What would that value be?) and the greatest dilution of Phenol that also kills in 10 minutes but not 5 minutes. These values will be used to solve the equation shown below:

$$PC = \frac{\text{reciprocal of the dilution of Disinfectant "X"}}{\text{reciprocal of the dilution of phenol}}$$

For the example given, the greatest dilution of Disinfectant "X" that kills in 10 minutes but not 5 minutes would be 1:250, and that for Phenol would be 1:50. Placing the reciprocals of these values (250 and 50, respectively) in the equation, we get

$$PC = \frac{250}{50}$$

or a phenol coefficient value of 5. What does this value mean? We know that Disinfectant "X" is 5 times more effective against *Staphylococcus aureus* than is phenol under the test conditions.

Table 23.2 gives the phenol coefficient value for several common antiseptics or disinfectants when tested against *Staphylococcus aureus* and *Salmonella typhi*—the standard test organisms.

Table 23.2   Phenol Coefficient Values

Chemical Agent	*Staphylococcus aureus*	*Salmonella typhi*
Phenol	1.0	1.0
Tincture of Iodine	6.3	5.8
Lysol	5.0	3.2
Mercury chloride	100.0	143.0
Ethyl alcohol	6.3	6.3
Hydrogen peroxide	—	0.01

CHOOSING AN IDEAL DISINFECTANT

As we have seen, there are many different types of disinfectant or antiseptic agents. When selecting one for use several criteria should be considered. An ideal disinfectant should be:

1. Highly effective against many organisms in dilute concentrations;

2. Fast-acting even in presence of organic material;

3. Non-toxic if applied topically, inhaled or ingested;

4. Stable;

5. Soluble in water or alcohol;

6. Able to penetrate material to be disinfected without damaging it;

7. Colorless, odorless;

8. Biodegradable;

9. Inexpensive and readily available.

No agent is likely to satisfy all of these criteria for the desired application. We must then choose the one agent that best satisfies our needs.

LABORATORY OBJECTIVES

In this exercise you will use the disk dilution technique to evaluate the effectiveness of several disinfectants and antiseptics. You should

- Distinguish between disinfectants and antiseptics.

- Distinguish between bactericidal and bacteriostatic action of chemical agents.

- Evaluate the effects of antiseptics or disinfectants on selected bacterial cultures.

- Discuss the limitations of using the disk-dilution method for evaluating chemical agents.

MATERIALS NEEDED FOR THIS LABORATORY

1. Twenty-four hour broth cultures of:
   Escherichia coli
   Staphylococcus aureus
   Bacillus subtilis

2. Media, per group:
   Mueller-Hinton agar plates (3)
   Blood agar plates (3)

3. Varied antiseptics and/or disinfectants. At least 6 should be provided for student use. Students may also bring in any antimicrobial agents they would like to test.

4. Petri dish containing sterile filter paper disks.

5. Sterile swabs.

6. Forceps.

PROCEDURE

1. Obtain 1 plate of Mueller-Hinton agar and 1 plate of Blood agar for each organism being tested. Label with the test organism and your identifying information.

2. Dip a sterile swab into a broth culture of the test organism. Swab the entire surface of the Mueller-Hinton and Blood agar plates.

3. Repeat steps 1 and 2 for each organism being used.

4. Determine which antiseptics or disinfectants you will use. The same 6 agents should be used

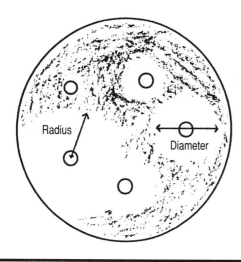

Figure 24.2   Measuring the Zone of Inhibition

on each of your plates. Indicate on the bottom of each plate the position for each of the disks and the chemical agent present on the disk.

5. Using your sterile forceps, pick up one of the sterile filter-paper disks. Dip the edge of the disk into the desired solution and allow it to absorb the solution by capillary action.

6. Place the disk in the designated area on the inoculated petri plate. Be sure to gently press down the disc on the agar surface using your forces. This will assure better adherence to the surface when the plates are inverted for incubation.

7. Repeat steps 5 and 6 for each of the test solutions on each of your plates, being sure to resterilize your forceps before each application.

8. Incubate your plates at 37°C until the next laboratory session.

9. Following incubation, the zones of inhibition should be determined for each solution on each plate. To determine the zones, measure either the diameter of the zone, or double the radius of the zone as measured in mm. (See **Figure 23.2**) Record your results on the Laboratory Report Form.

Name _____

Section _____

LABORATORY REPORT FORM

EXERCISE 23
ASSAY OF ANTIMICROBIAL AGENTS: DISK-DIFFUSION METHODS

What was the purpose of this exercise?

1. For each of the disinfectants or antiseptics tested, determine the active ingredient. What class of inhibitory agent does it belong to and what is its mode of action?

CHEMICAL AGENT	ACTIVE INGREDIENT(S)	CLASS OF AGENT	MODE OF ACTION

2. Zones of inhibition (in mm):

CHEMICAL AGENT	Escherichia coli		Staphylococcus aureus		Bacillus subtilis	
	Mueller-Hinton agar	Blood agar	Mueller-Hinton agar	Blood agar	Mueller-Hinton agar	Blood agar

3. Why was Blood agar used in addition to Mueller-Hinton agar?

4. What variations in results did you observe when comparing effectiveness of the tested solutions for:

    a. the Blood agar plate and the Mueller-Hinton plate for the same test organism?

    b. the varied organisms on the Mueller-Hinton plates?

QUESTIONS

1. What factors, other than the agent being tested, can influence the size of the zone of inhibition?

2. Were any colonies observed within the zone of inhibition? Explain why such colonies might be observed?

3. The disk-diffusion technique does not enable us to distinguish between bacteriostatic and bactericidal effects of chemical agents. Design an experiment that would enable you to determine whether the organisms were inhitibed or killed.

ASSAY OF ANTIMICROBIAL AGENTS:
USE-DILUTION METHOD

We saw in Exercise 23 that the overall effectiveness of antimicrobial agents can be determined by using a disk-diffusion method. What could not, however, be determined by this method was whether the microorganisms were killed or simply inhibited. In this exercise we will determine the bactericidal or bacteriostatic nature of select antimicrobial agents.

BACKGROUND

If antimicrobial agents are bactericidal, they will irreversibly damage the cell so that the cell is no longer viable, whether the chemical agent is present or not. This reaction cannot be reversed. Bacteriostatic agents, on the other hand, are only effective when they are in direct contact with the microorganism. When the chemical agent is removed, the organism is again able to grow. In other words, the effect is reversible.

To prove bactericidal activity, the agent must be removed from contact by either washing the cells in saline or by diluting the agent down to a concentration that is no longer effective. Cells which have been irreversibly damaged by the antimicrobial agent will not be able to grow when inoculated into a suitable growth medium in the absence of the agent.

STANDARD USE-DILUTION TEST

The standard method of testing for use-dilution, is the American Official Analytical Chemist's use-dilution test. In this test, chemicals are tested against three different bacterial species—*Salmonella cholera-suis, Staphylococcus aureus,* and *Pseudomonas aeruginosa*. Metal rings are dipped into standardized cultures, removed and dried. They are then placed into the recommended concentrations (the concentration of use) of the disinfecting agent for 10 minutes at room temperature, after which they are transferred to appropriate growth media. Following incubation, tubes are inspected for signs of microbial growth. Any growth would indicate bacteriostasis, and the absence of growth would indicate bactericidal activity.

LIMITATIONS TO THE STANDARD TEST PROCEDURE

The standard test as outlined above can effectively demonstrate the bactericidal or bacteriostatic effect of a chemical agent, but it does not show how other factors can influence the antimicrobial effect of these chemicals. We saw in Exercise 23 that the presence of organic material can influence the effectiveness of an antimicrobial agent. Other important factors would include the penetration capabilities of the chemical agent and the time of exposure.

We will be performing a modification of the standard use-dilution test. Toothpicks and stainless steel pins will be soaked in a bacterial suspension and dried. They will then be placed in one of the antimicrobial agents used in Exercise 23 for varied periods of time. The stainless steel pins are comparable to the disinfection of a non-porous surface, whereas the toothpicks will represent a porous surface.

Following exposure, the pins and toothpicks will be transferred to tubes of nutrient broth and incubated so that any viable bacteria will grow. The volume of the broth in the tubes (approximately 10 ml) is so large relative to the amount of chemical agent adhering to the surface of the pin or toothpick that they will be diluted to a non-effective concentration. If no growth is observed following incubation, the bacteria can be assumed to have been killed and the agent said to be bactericidal. If growth occurs in the tubes, the agent is bacteriostatic at best. Is it possible to have a zone of inhibition with the disk-diffusion test and growth with the use-dilution test? What about no zone of inhibition with the disk-diffusion test and no growth in the use-dilution test? What could you conclude if there was both a zone of inhibition and no growth? No zone of inhibition and growth?

OBJECTIVES

In this exercise you will be performing a modified use-dilution test using an antimicrobial agent that was also tested by the disk-diffusion method. Following the completion of this exercise you should:

- Understand the difference between bacteriostatic and bactericidal agents.

- Explain how to test to determine whether chemicals are bacteriostatic or bactericidal.

. • Describe the use-dilution test.

- Discuss those factors which could influence the results of a use-dilution test.

MATERIALS NEEDED
FOR THIS LABORATORY

1. Broth cultures of the following:
   Escherichia coli (24 hours)
   Micrococcus luteus (24 hours)
   Bacillus subtilis (72 hours)

2. One of the antimicrobial agents used in Exercise 23.

3. Two empty sterile petri plates.

4. Six toothpicks and six stainless steel pins.

5. Twelve tubes of sterile nutrient broth (approximately 10 ml per tube).

6. Filter paper.

7. Forceps.

PROCEDURE

NOTE: Each group will be assigned to use one of the microorganisms for the test procedure.

Figure 24.1 provides a flow chart for this procedure. Be sure to carefully examine it before starting.

1. Divide the nutrient broth tubes into two sets—one for use with the toothpicks and one with the stainless steel pins.

2. Label each set of tubes as follows:
   #1—toothpick, O time (control)
   #2—toothpick, 1 minute soak
   #3—toothpick, 2 minute soak
   #4—toothpick, 5 minute soak
   #5—toothpick, 10 minute soak
   #6—toothpick, 15 minute soak
   #7—pin, 0 time (control)
   #8—pin, 1 minute soak
   #9—pin, 2 minute soak
   #10—pin, 5 minute soak
   #11—pin, 10 minute soak
   #12—pin, 15 minute soak

3. Pour 10 ml of the assigned bacterial culture into one of the petri dishes. Add the toothpicks and stainless steel pins and allow them to soak for about 5 minutes. Be sure that they are covered by the broth suspension.
   NOTE: When making the transfers into your nutrient broth tubes, drop the toothpick or pin into the broth. Do not allow the forceps to touch the broth in the tubes. Be sure to sterilize your forceps between each use.

4. Using your forceps, remove the toothpicks and pins from the broth and place them on a piece of filter paper to eliminate any excess of broth culture.

5. Transfer one toothpick into tube #1 and one pin into tube #7. These are your "zero time" or control tubes. What is the purpose of using these controls?

6. Pour 10 ml of the antimicrobial agent into the second petri dish. Add the remaining 5 toothpicks and 5 pins to this dish. Start timing as soon as they are added.

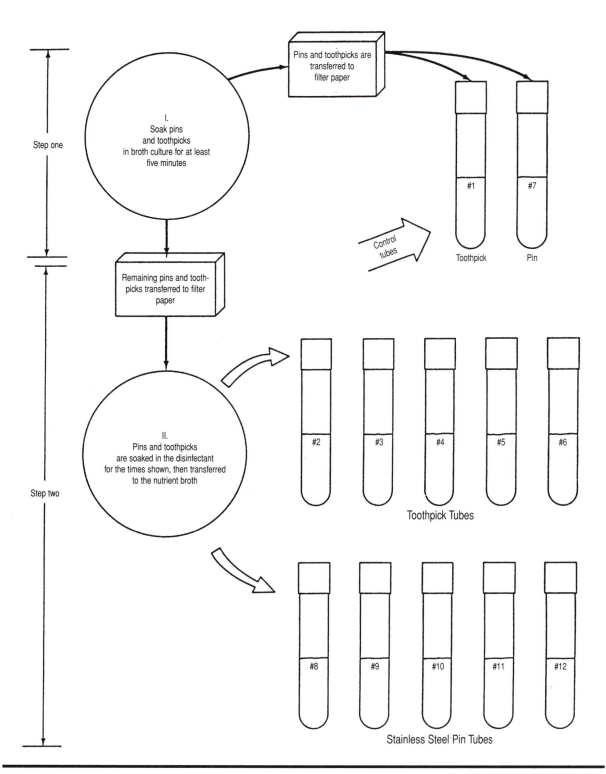

Figure 24.1   Use-dilution Assay Procedure

7. Transfer one toothpick and one pin to the appropriate nutrient broth tubes at the desig-

TIME	Place Toothpick into Tube #	Place Pin into Tube #
1 minute		
2 minutes		
5 minutes		
10 minutes		
15 minutes		

nated time intervals. Remember the time is determined from the minute they were placed into the antimicrobial agent.

8. Place all tubes in the incubator.

9. Be sure to carefully dispose of the petri plates as indicated by your instructor.

10. Following incubation, record whether or not growth occurred in each tube. Compare the results for your organism with those for the other organisms. Record the results on your Laboratory Report Form.

Name _____

Section _____

LABORATORY REPORT FORM

EXERCISE 24
ASSAY OF ANTIMICROBIAL AGENTS: USE-DILUTION METHODS

What was the purpose of this exercise?

1. Antimicrobial agent tested: _____

2. What are the active ingredient(s) in the antimicrobial tested?

3. Complete the following table. Indicate any growth as (+).

	Escherichia coli		*Micrococcus luteus*		*Bacillus subtilis*	
TIME	Toothpick	Pin	Toothpick	Pin	Toothpick	Pin
0						
1 min.						
2 min.						
5 min.						
10 min.						
15 min.						

4. Would you consider the antimicrobial agent tested to be bacteriostatic or bactericidal? Explain your conclusion.

5. Could you determine from these results if the agent is sporicidal? Explain why or why not.

6. Summarize the results achieved by other groups for other antimicrobial agents tested.

Chemical Agent	Bactericidal	Bacteriostatic

QUESTIONS:

1. Explain any differences that might be observed between the killing times noted for the pins and toothpicks.

2. What test(s), besides the use-dilution test, would you need to use to determine the effectiveness of an antiseptic agent?

3. What factors would contribute to varied results for *Micrococcus luteus* and *Escherichia coli*? For *Bacillus subtilis* and *Micrococcus luteus*?

ANTIBIOTIC SENSITIVITY TESTING:
THE KIRBY-BAUER PROCEDURE

Successful isolation and identification of clinically significant bacteria from a clinical specimen is only half of the task expected of the microbiology laboratory. Virtually every isolate that may have clinical significance must be tested to determine its antibiotic sensitivity pattern.

There are two reasons for this: First, the physician needs to have the sensitivity pattern available to protect the patient against the possibility that this particular isolate may be resistant to the commonly used antibiotics. Second, antibiotic sensitivity patterns have proven to be consistent for a given organism and can be used as an additional diagnostic characteristic.

BACKGROUND

Alexander Fleming first noted in 1929 that *Penicillium* growing on an agar plate could inhibit the growth of *Staphylococcus aureus.* This was followed by Selman Waksman discovering that *Streptomyces griseus* produced streptomycin which inhibited many Gram-negative organisms. The additional discovery in the 1930s that sulfanilamide would also inhibit certain infectious agents began the era of chemotherapy and our ability to cure microbial infections.

Today we have a large number of antimicrobial agents available for treating infections. These have become an essential part of modern medical practice by using selective toxicity to eliminate infecting organisms or preventing the establishment of infection.

Antibiotics, derived from the term antibios (against life), were originally defined as those chemicals, produced by microorganisms, that inhibit the growth of other microorganisms. This included only those naturally occurring agents, making a distinction between them and the synthetic chemical agents, such as the sulfonamides. Today we find that the distinction between the natural and synthetic drugs has pretty much disappeared as synthetic versions of naturally occurring agents are commonly used. Consequently, we find most of those antibacterial drugs used in the treatment of infection being referred to as either antibiotics or chemotherapeutic agents.

Antibiotics that are effective against a wide range of organisms, including both Gram-positive and Gram-negative species, are referred to as being *broad-spectrum. Narrow-spectrum* drugs, on the other hand, are effective against a very select group of organisms. In all cases, the use of antibiotics is dependent upon the exhibition of *selective toxicity.* This is achieved by involving biochemical features of the pathogen that are not possessed by the host cells, therefore making the antibiotic toxic to the microorganism but not to the host. The five major modes of action utilized by antimicrobial drugs include (1) the inhibition of cell wall synthesis, (2) the disruption of cell membrane functioning, (3) the inhibition of protein synthesis, (4) the inhibition of nucleic acid synthesis, and (5) competitive inhibition of metabolic pathways. See your text book for further discussion of these modes of action.

An emerging problem today is the rapid development of microorganisms resistant to the effects of chemotherapeutic agents. The following factors have been cited for contributing to this problem:

- development of alternative biochemical pathways by microorganisms, circumventing the effect of the chemotherapeutic agent.

- production of enzymes by microorganisms which inactivate the chemotherapeutic agent.

- presence of drug-resistant plasmids which provide resistance to multiple drugs for some bacteria.

- undertreatment of infections resulting in the survival of the most resistant strains.

- overuse of antibiotics resulting in the increased prevalence of resistant organisms through the wide-spread destruction of susceptible strains.

PPNG (penicillinase-producing *Neisseria gonorrhoea*) and MRSA (methicillin-resistant *Staphylococcus aureus*) are just two of the resistant forms being frequently encountered today. We will continue to see more species becoming resistant and the reemergence of pathogens previously considered "under control" (i.e., *Mycobacterium tuberculosis*) if the overuse and unnecessary use of antibiotics continues.

Today, the need for careful testing of pathogens for drug sensitivity is becoming more essential. There are two major tests that are used to determine the susceptibility of organisms to chemotherapeutic agents—the Minimum Inhibitory Concentration (MIC) test and the Kirby-Bauer disk-diffusion test.

MINIMUM INHIBITORY CONCENTRATION (MIC) TEST

A precise way to measure the sensitivity of an organism to a given antibiotic is determining the minimal inhibitory concentration (MIC) of an antibiotic. The MIC value is the lowest concentration of an antibiotic that inhibits the growth of a given bacterial species under specified conditions. It is used to establish the concentration of antibiotic that must be maintained in the body to effectively inhibit the growth of the pathogen. It can also be used to determine the degree of drug resistance exhibited by a species.

To determine the MIC value for a particular antibiotic and a given microbial species, a standardized inoculum is added to tubes containing serial dilutions of the antibiotic. The tubes are examined for signs of turbidity which would indicate growth of the organism. (See **Figure 25.1**) Today microtiter plates and automated inoculation and reading systems are commonly employed.

The MIC test is not routinely performed in clinical laboratories due to the time and expense involved. It does, however, have the advantage in being able to be used with normally sterile body fluids, such as cerebrospinal fluid and blood, without the prior isolation of the infective agent(s).

THE KIRBY-BAUER DISK DIFFUSION TEST

Diffusion tests have been used throughout this century to determine the effect of various chemical agents on bacterial growth. Originally wells were cut in agar plates and the chemical agent was placed there. This was replaced by the use of paper disks containing the chemical agent following their introduction in 1947 by Bondi. Disk-diffusion tests, as they became known, had little credibility, however, until Bauer, Kirby, Sherris, and Turck developed, in 1966, the correlations between the MIC results for a large number of bacterial strains and the results achieved by a standardized disk susceptibility test. The Kirby-Bauer test has become the standard method for testing for antibiotic susceptibility in clinical laboratories. By correlating the observed zones of inhibition with the known MIC values, they were able to standardize zone interpretations as indicating *susceptibility* (S) or *resistance* (R) by the given organism. If the zone size did not correlate well with the MIC standards, the results were considered *inconclusive* (I). Table 25.1 shows the correlation between zone of inhibition and susceptibility for some of the antibiotics commonly used.

The Kirby-Bauer procedure is precisely controlled to maintain the validity of the test results. The controls include:

Standardized bacterial suspension. The amount of antibiotic needed to inhibit a population of bacteria is proportional to the size of the population. The density of the bacterial suspension used in the Kirby-Bauer procedure is maintained constant by comparing it to a standardized suspension of a known turbidity. The standard used is the McFarland Standard Tube No. 0.5.

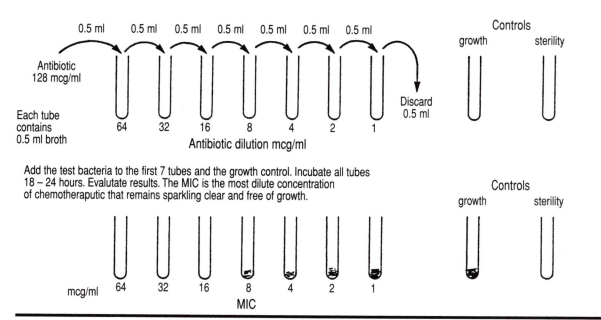

Figure 25.1   Minimum Inhibitory Concentration (MIC) Test

Standardized concentrations of antibiotics in the disks. All other things being the same, the diffusion rate of any chemical is directly proportional to the concentration of that chemical. The amount of antibiotic in the disk must be standardized if any consistency in the diameters of the zones is to be expected. Standardized disks are available from biological supply houses.

Standardized medium. The Kirby-Bauer procedure uses a medium known as Mueller-Hinton agar. This medium was chosen because its properties relative to diffusion of antibiotics and growth of bacteria are known. Any other medium would not necessarily allow the antibiotic to diffuse at the same rate and would not support the same amount of bacterial growth. The depth of the medium must also be standardized. (Why would the depth of the medium affect the size of the zone of inhibition observed?) The accepted depth of the medium is 4.0 mm.

Standardized incubation time and temperature. The rate of diffusion is temperature dependent and increases as the temperature is increased. Furthermore, if the time for diffusion is extended indefinitely, the concentration of the antibiotic in the medium will eventually be equal everywhere. Antibiotics are bacteriostatic, and if the concentration falls below the minimum inhibitory concentration (MIC) the bacteria will resume growth. An incubation temperature of 35 to 37°C for no more than 24 hours (usually 16 – 18 hours) is required.

Following incubation, the diameter of the zones of inhibition are measured and this value is compared with the standard table for interpretation as indicating susceptibility, resistance, or inconclusive (intermediate) results.

Sensitive. The minimum zone diameter, expressed in millimeters, that indicates the organism will be sensitive to the antibiotic at normal therapeutic doses. If a zone diameter larger than this value is observed, the organism is considered to be sensitive to the antibiotic.

Resistant. The minimum zone diameter, expressed in millimeters, that indicates the organism will be resistant to the antibiotic at normal therapeutic doses. If a zone diameter smaller than this value is observed, the organism is considered to be resistant to the antibiotic.

Inconclusive or Intermediate. The range of zone diameter, expressed in millimeters, between sensitive and resistant zone diameters where individual differences between patients and bacterial strains may cause one particular combination to be resistant, while another combination might prove to be sensitive.

Table 25.1   Interpretive Standards for Disk-Diffusion Suceptibility Testing

ZONE OF INHIBITION (mm)				
ANTIMICROBIAL AGENT	DISK CONTENT	RESISTANT	INTERMEDIATE	SUSCEPTIBLE

Ampicillin
  Enterobacteriaceae
  Staphylococci
  Enterococci
  Streptococci
	10 mcg	≤ 13
		≤ 28
		≤ 16
		≤ 21
			14 – 16
			
			22 – 29
				≥ 17
				≥ 29
				≥ 17
				≥ 30

Bacitracin	10 units	≤ 8	9 – 12	≥ 13
Cephalothin	30 mcg	≤ 14	15 – 17	≥ 18
Chloramphenicol	30 mcg	≤ 12	13 – 17	≥ 18
Erythromycin	15 mcg	≤ 13	14 – 22	≥ 23
Gentamicin	10 mcg	≤ 12	13 – 14	≥ 15
Kanamycin	30 mcg	≤ 13	14 – 17	≥ 18
Methicillin Staphylococci	5 mcg	≤ 9	10 – 13	≥ 14
Neomycin	30 mcg	≤ 12	13 – 16	≥ 17
Nitrofurantoin	300 mcg	≤ 14	15 – 16	≥ 17

Penicillin G
  Staphylococci
  Enterococci
  Streptococci
	10 units	≤ 28
		≤ 14
		≤ 19
			20 – 27	≥ 29
				≥ 15
				≥ 28

Polymixin B	300 units	≤ 8	9 – 11	≥ 12
Streptomycin	10 mcg	≤ 12	12 – 14	≥ 14
Sulfisoxazole (Gantrisin)	250 mcg	≤ 12	12 – 16	≥ 17
Sulfonamides	300 mcg	≤ 12	13 – 16	≥ 17
Tetracycline	30 mcg	≤ 14	15 – 18	≥ 19

Examine Table 25.1. Note that for some of the antibiotics (ampicillin and penicillin G) the proper evaluation of the zones is dependent upon the organism that is being tested. Why do you think there are higher values required for some organisms than others?

If any growth of colonies occurs within the zone of inhibition, they must be Gram stained to determine if they are contaminants or a mutant (resistant) strain of the pathogen.

The data determined by the Kirby-Bauer test provides the physician with a screening test to determine which drugs are exhibiting significant antibiotic activity. Final decision on the antibiotic of choice is not, however, dependent upon that drug which produces the largest zone of inhibition. Such additional factors as potential of side effects, biotransformation potential (whether the antimicrobial agent will remain active in the body long enough to be selectively toxic to the pathogen), and the ability of the antimicrobial agent to reach the site of the infection in adequate concentration must all be taken into account.

OBJECTIVES

In this exercise you will be determining the antibiotic-sensitivity pattern for several organisms. In performing this test you will:

- Perform the Kirby-Bauer test for the determination of antibiotic location of the disks.

- Differentiate between the MIC and the Kirby-Bauer tests of antibiotic susceptibility.

- Distinguish between broad-spectrum and narrow-spectrum antibiotics.

- Determine the presence of drug-resistant organisms.

- Interpret the results of the antibiotic sensitivity test based on the zone size.

- Discuss the medical importance of the antibiotic-sensitivity test.

MATERIALS NEEDED FOR THIS LABORATORY

1. Three Mueller-Hinton agar plates poured to a depth of 4.0 mm.

2. Broth cultures of:
   Staphylococcus aureus
   Escherichia coli
   Pseudomonas aeruginosa

3. Sterile swabs

4. Antibiotic discs for eight different antibiotics in either single disk dispensers or a multiple dispenser (See **Figure 25.2**)

5. Forceps

LABORATORY PROCEDURE

1. Obtain and label one Mueller-Hinton plate for each organism being tested.

2. Note the code present on the antibiotic disks. You do not need to label your plates for the location of the disks.

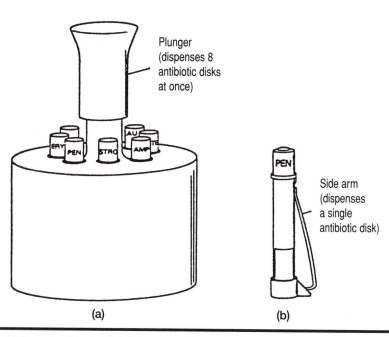

Figure 25.2   Antibiotic Disk Dispensers: (a) Multiple Dispenser; (b) Single Disk Dispenser.

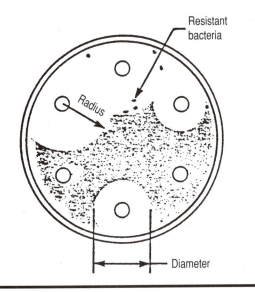

Figure 25.3   Determining Zones of Inhibition

3. Swab the surface of each plate with the organism being tested. Be careful to cover the entire surface of the plate so that you will get a solid, continuous lawn of bacteria. Dispose of your swab as indicated by your instructor.

4. Place the antibiotic disks on the plates.

    a. If a multiple dispenser is used, place it over the plate and slowly depress the plunger. Using your sterile forceps, press lightly on each disk to be sure that it will adhere to the surface of the medium when the plate is inverted. Repeat for each plate.

    b. If using single disk dispensers, carefully apply each of the disks to your plates being sure to keep them equally spaced. Using your sterile forceps, press lightly on each disk to be sure that it will adhere to the surface of the medium when the plate is inverted. Repeat for each plate.

5. Incubate the plates for 24 hours. If you are unable to read the results at that time, they should be refrigerated until the next laboratory period.

6. Measure (in mm) the zone of inhibition for each antibiotic.

7. Record your results on the Laboratory Report Form. Be sure to note the presence of any colonies within the zone of inhibition.

LABORATORY REPORT FORM

EXERCISE 25
ANTIBIOTIC SENSITIVITY TESTING: THE KIRBY-BAUER PROCEDURE

What was the purpose of this exercise?

1. Complete the following chart:

Chemotherapeutic Agent	Disk Code	Staphylococcus aureus		Escherichia coli		Pseudomonas aeruginosa	
		Zone (mm)	S, I, or R	Zone (mm)	S, I, or R	Zone (mm)	S, I, or R

2. Which of the chemotherapeutic agents tested were, based on your results, broad-spectrum? Narrow spectrum?

3. Were any colonies growing within any of your zones of inhibition? If so, how would you determine if it was a contaminant or a resistant colony?

QUESTIONS:

1. What do the terms PPNG and MRSA mean?

2. What advantage is there to performing the MIC tst? the Kirby-Bauer test?

3. What is the significance of colonies that develop within otherwise clear zones of inhibition?
    If the laboratory report for one of your patients indicated colonies within the zone, what
    concerns would you have for your patient?

4. Why is *Pseudomonas aeruginosa* of such concern in burn patients and in immunologically
    compromised patients?

5. A culture is inoculated into the following series of test tubes containing the designated concentra-
    tion of an antibiotic and nutrient broth. After a 24 hour incubation at 37°C, the following results
    are observed. What would be reported as the MIC value for the test organism and antibiotic?

MIC

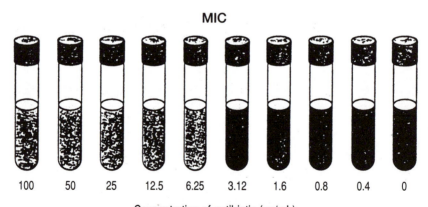

| 100 | 50 | 25 | 12.5 | 6.25 | 3.12 | 1.6 | 0.8 | 0.4 | 0 |

Concentration of antibiotic (µg/mL)

DETECTION OF MUTANT STRAINS OF BACTERIA

In this exercise we are going to explore several methods of identifying the presence of genetic variations in bacteria. These variations may be either induced (due to exposure to a known mutagenic agent) or spontaneous (naturally occurring without any known exposure to a mutagenic agent).

BACKGROUND

Mutations are changes in the base sequence of DNA that are transmitted to daughter cells and which may result in an inheritable change in the phenotype of a cell. Spontaneous mutations occur at a relatively low rate, varying from 1 in 10^4 to 1 in 10^{12} cell divisions. When we consider the rapid rate of cell division that occurs in bacteria, we can see that there are likely to be a number of mutants found in any bacterial population. For example, one gene present in *Escherichia coli* is the Gal + gene. This gene enables the organism to metabolize galactose. It will undergo a spontaneous mutation to the Gal + form, which is unable to utilize galactose, at a rate of approximately 2 in 10^7 cell divisions.

If the new phenotype (visible trait) is selected against by the environment, the cell will disappear from the population. If, on the other hand, the change is selected for, the new cell type may gradually increase in number and eventually may become a significant part of the population. A "neutral" mutation—one that is neither selected for nor against—will have little, if any, effect on the distribution of the mutant in the population.

The development of a mutant population, then, is dependent upon two processes: the mutation itself and the selection of that mutant by the environment. An example of this would be the development of drug resistance in *Staphylococcus aureus*. The ability to synthesize penicillinase, an enzyme which inactivates most penicillins, was of no particular advantage to *Staphylococcus aureus* before penicillin became a widely used antibiotic. Penicillin did not become the well-known "wonder drug" until during World War II when it was used extensively to treat wounds. After the war it became readily available to the public and was widely (indiscriminately?) used to treat virtually any bacterial infection and even viral infections such as a bad cold.

Nobody knows when the mutation for penicillin resistance first appeared in the population of *S. aureus*. Extensive selection for penicillinase-positive bacteria could not occur as long as penicillin was not widely distributed. As the antibiotic came into greater usage, the mutant (drug-resistant) strain was suddenly placed at an advantage. It could inactivate the antibiotic, while the nonmutant (wild type) cells could not. Selection for the resistant members of the population was so extensive that it is now relatively uncommon to encounter an isolate of *S. aureus* that is sensitive to the original penicillin G. Due to the continued overuse of antibiotics, we are currently seeing a significant amount of drug resistance in *Pseudomonas aeruginosa* and *Mycobacterium tuberculosis* infections in addition to that discussed in *Staphylococcus aureus*.

If a mutation results in the inability of an organism to synthesize a needed enzyme, it will have a nonfunctional metabolic pathway. If the pathway effected is for the production of a substance essential for the growth of the cell, the cell will die unless the needed substance is provided in its medium. These are considered nutritional mutants.

Mutant strains cannot be detected unless we expose the bacteria to an environment that would select for that mutant. How would you detect penicillin-resistant bacteria unless you exposed the population to penicillin? While this seems selfevident, it does raise an interesting question: How can you prove that the mutation occurs independently from its selection?

The selection of mutant cells can be by **direct selection,** where only those cells exhibiting the mutation can grow, or by **indirect selection.** An example of direct selection would be the addition of an antibiotic to the growth media for an organism which is normally inhibited by the antibiotic. Any colonies that grow would be due to the presence of the mutation. In indirect selection bacteria are grown in an environment that is not selective for a given mutation. A replica plate is used to transfer members of each colony to a selective environment. By comparing the presence of colonies following incubation we can indirectly determine the mutated colonies by their absence in the selective environment.

In this laboratory exercise, we will utilize both direct and indirect selection of mutated colonies. We will also be detecting the presence of spontaneous mutations and the induction of induced mutations.

OBJECTIVES

Bacterial mutations appear randomly in the population at a predictable rate. The mutation rate can be increased with certain mutagenic agents, but the resulting mutations are still randomly determined. A mutant population must be selected for by applying the correct environmental conditions. You should:

- Understand the relationship between mutation and selection.

- Understand that mutation rates can be altered by exposure to mutagenic agents, but that the randomness of these mutations cannot be altered.

- Understand the relationship between the indiscriminate use of antibiotics and the development of resistant strains of bacteria.

- Distinguish between direct and indirect selection of mutations.

- Be able to use the replica plating technique.

MATERIALS NEEDED FOR THIS LABORATORY

A. SPONTANEOUS MUTATION DETECTION

1. Four nutrient agar tails.

2. 50°C water bath.

3. Antibiotic stock solutions. The stock solution should be prepared so that when 1 ml is added to a nutrient agar tall it will exceed the minimal

Antibiotic	Stock Solution Concentration	Medium Concentration
Penicillin	1000 mcg/ml	50.0 mcg/ml*
Streptomycin	1000 mcg/ml	50.0 mcg/ml*
Ampicillin	500 mcg/ml	25.0 mcg/ml*
Polymyxin B	500 mcg/ml	25.0 mcg/ml*
Erythromycin	1000 mcg/ml	50.0 mcg/ml*
Gentamicin	100 mcg/ml	5.0 mcg/ml*
Tetracycline	250 mcg/ml	12.5 mcg/ml*
Chloramphenicol	500 mcg/ml	25.0 mcg/ml*
Kanamycin	250 mcg/ml	12.5 mcg/ml*

*These values based on estimated in vivo dosage. Other values based on published MIC data.

inhibitory concentration (MIC) for the organism you are testing. You will use streptomycin and two other antibiotics.

4. One tube of sterile water.

5. Nutrient broth suspension of one of the following:
   Serratia marcescens
   Escherichia coli

6. Sterile 1.0 ml pipettes (5)

7. Sterile petri plates (4)

8. Bent glass spreader

9. 95% ethanol for sterilizing bent glass spreader.

10. Beaker

B. INDUCED MUTATION DETECTION

1. Nutrient agar plates (4)

2. Minimal Salts with glucose agar plate (1)

3. Nutrient broth suspension of one of the following:
   Serratia marcescens
   Escherichia coli

4. 99-ml sterile water dilution blanks (3)

5. Sterile 1.0 ml pipettes

6. Bent glass spreader

7. 95% ethanol for sterilizing bent glass spreader

8. UV light

9. Replica plate apparatus

10. Sterile velveteen square

11. Rubber band

LABORATORY PROCEDURE

A. SPONTANEOUS MUTATION DETECTION

1. Obtain four sterile petri plates. Label three of them with the antibiotic, its concentration, and the name of the organism being tested. The fourth plate should be labeled "Control."

2. Melt tubes of nutrient agar and cool in water bath to 50°C.

3. Obtain one tube of melted agar. Aseptically add 1.0 ml of the streptomycin stock solution to the tube, mix, pour the medium into the appropriately labeled petri plate, and allow it to solidify.

NOTE: When sterilizing the bent glass spreader in the next step, be sure that you do not hold the rod over anything flammable. Ignited alcohol may drop off the rod and set fire to whatever it lands on. Clear the countertop of all unnecessary materials.

4. Repeat step 3 with two more antibiotic solutions.

5. The control is prepared by aseptically adding 1.0 ml of sterile water to a tube of medium,

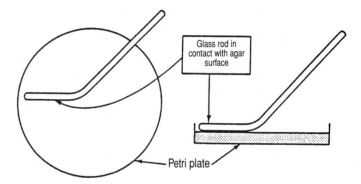

Figure 26.1   Use of Bent Glass Spreader

mixing, pouring it into the petri plate labeled "Control," and letting it solidify.

6. Aseptically transfer 0.1 ml of the bacterial suspension to each of the agar plates.

7. Sterilize the bent glass spreader by dipping it in 95% ethanol, igniting the alcohol in the Bunsen burner flame, and burning the alcohol off. Let the spreader cool.

8. Quickly spread the suspension over the surface with the sterilized bent glass spreader (See **Figure 26.1**)

9. Incubate the *E. coli* plates at 37°C and the *Serratia marcescens* plates at room temperature for 24 hours. If the results cannot be read at that time, refrigerate the plates until the next laboratory period.

10. Count the colonies that develop. How can you determine that they represent antibiotic-resistant colonies of your test organism and are not contaminants?

11. Record your results on the Laboratory Report Form, Part A.

B. INDUCED MUTATION DETECTION

1. Obtain three nutrient agar plates. Label each with the organism used and the dilution (1:10,000, 1:100,000 or 1:1,000,000).

2. Label your sterile water blanks "1," "2," and "3."

3. Prepare a dilution series of your organism. Refer to Exercise 8, and to **Figure 26.2**

    a. Aseptically transfer 1.0 ml of your broth suspension to the first 99 ml sterile water blank. Mix it well. (What is the dilution in this bottle?)

    b. Using another pipet, transfer 1.0 ml from water blank "1" to sterile water blank "2". Mix it well. (What dilution do you now have?)

    c. Using another pipet, transfer 1.0 ml from water blank "2" to sterile water blank "3." Using the same pipet, also transfer 1.0 ml from water blank "2" to the surface of the 1: 10,000 plate and 0.1 ml to the surface of the 1: 1 00,000 plate.

    d. Mix water blank "3" and transfer 1.0 ml to the surface of the 1 :1 ,000,000 plate.

4. Sterilize the bent glass spreader by dipping it in 95% ethanol, igniting the alcohol in the Bunsen burner flame, and burning the alcohol off. Let the spreader cool.

5. Spread the dilution over the surface of the plate with the bent glass spreader. Repeat with the remaining plates.

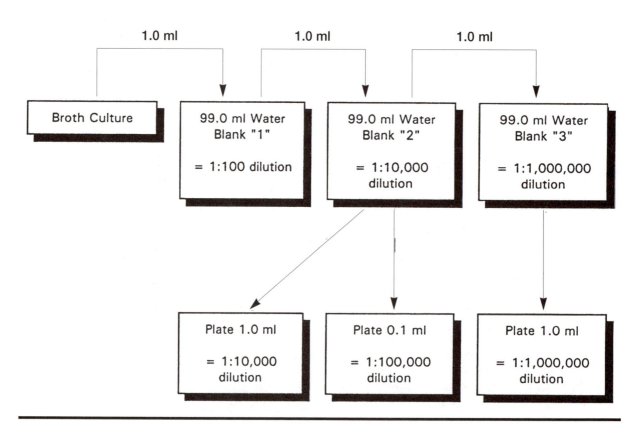

Figure 26.2   Preparation of Dilution Series

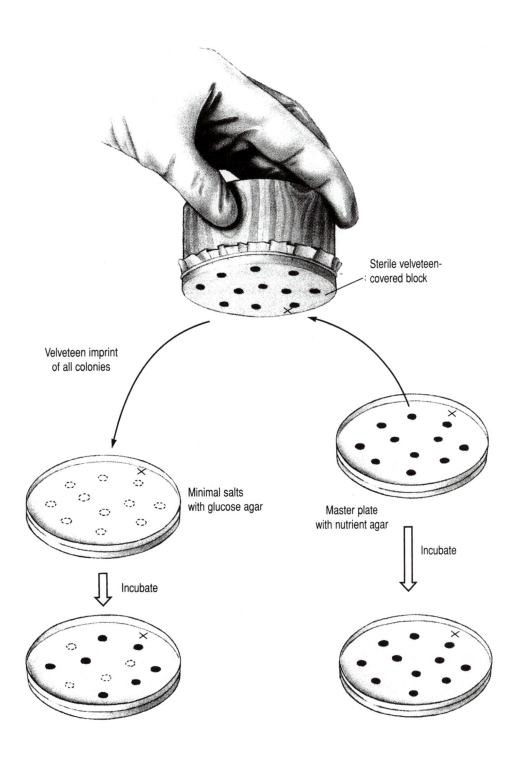

Sterile velveteen-covered block

Velveteen imprint of all colonies

Minimal salts with glucose agar

Master plate with nutrient agar

Incubate

Incubate

Figure 26.3　　Use of Replica Plate Apparatus

6. Place each of your inoculated nutrient agar plates under the UV light for 30 seconds.

7. Invert the plates and incubate until the next laboratory period. The *Serratia marcescens* should be incubated at room temperature, the *E. coli* at 37°C.

8. Determine the plate with 30 to 100 isolated colonies. This is your master plate that you will be using for the remainder of the exercise. Place a reference mark on the bottom of the plate (See **Figure 26.3**)

9. Obtain a plate of nutrient agar and of minimal salts with glucose agar. Label them and place a reference mark on the bottom of each.

10. Get a replica plating block. Place a piece of sterile velveteen cloth over the top. Touching only the corners of the cloth, bend it down over the surface of the block. Secure the cloth with a rubber band.

11. Inoculate your sterile agar plates in one of the following manners:

    a. Place the master plate and the uninoculated plates on the table with the reference marks all at 12 o'clock. Remove the covers. Carefully press the replica plate block to the surface of the master plate, aligning the handle with the reference mark. Without altering the position of the replica plate, touch it to the surface of the minimal salts with glucose agar plate and then to the nutrient agar plate. You should not go back to the master plate between inoculating your plates.

    b. Hold the replica plate device on the table with the velveteen surface up. Remove the lid from the master plate and invert it over the block, allowing the agar surface to lightly touch the velveteen surface. Replace the cover. Remove the cover from the minimal salts with glucose plate. Align the reference mark and lightly touch the surface of the medium on the inoculated velveteen surface. Replace the cover. Repeat ith the nutrient agar plate.

12. Place the velveteen square in an autoclavable container as indicated by your instructor.

13. Incubate the plates as before. Refrigerate the master plate.

14. Following incubation, compare the presence of colonies on the master plate, nutrient agar plate and minimal salts with glucose plate.

15. Record your results on the Laboratory Report Form, Part B.

LABORATORY REPORT FORM

EXERCISE 26
DETECTION OF MUTANT STRAINS OF BACTERIA

What was the purpose of this exercise?

A. SPONTANEOUS MUTATION DETECTION

Organism used: _____

Describe the growth present on the "Control" plate.

Complete the following table:

ANTIBIOTIC	CONCENTRATION IN MEDIUM	NUMBER OF RESISTANT COLONIES

How can you determine that the colonies seen are resistant mutants and not contaminants?

B. INDUCED MUTATION

Organism used: _____

Were any changes present in the appearance of the colonies following exposure to the UV light?

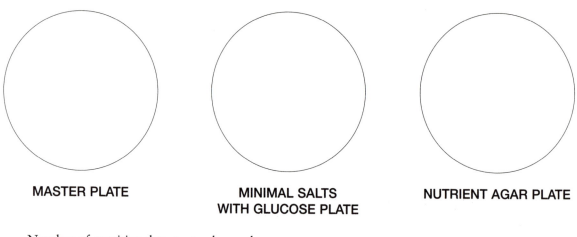

MASTER PLATE   **MINIMAL SALTS**   **NUTRIENT AGAR PLATE**
                          WITH GLUCOSE PLATE

Number of nutritional mutants observed: _____

QUESTIONS:

1. What environmental factors would contribute to the selection of spontaneous mutations?

2. What environmental factors would contribute to the induction of mutations in organisms?

3. Why did we use both the nutrient agar and the minimal salts plus glucose medium for the replica plating?

4. Distinguish between direct and indirect selection of mutations.

EPIDEMIOLOGY

For a pathogen to cause disease, it must come in contact with a susceptible host. This requires a method of transmission and a suitable portal of entry for the organism to gain access to the host. The study of the occurrence, distribution and control of these infectious diseases, as well as non-infectious diseases, is the science of *epidemiology*. Involved is not only determining the causative agent of the disease, but how it was transmitted, and how this spread can be slowed or stopped. In this exercise we will be simulating an epidemic with a known method of transmission.

BACKGROUND

Communicable diseases are characterized by their spread or transmission from person to person. The method of transmission may be by direct contact (sexually transmitted diseases), aerosol (influenza virus), water or food (cholera or salmonella), fomites (inanimate objects, such as a dirty nail, which can transfer a pathogen to a host), or vectors (a living organism, such as a fly or mosquito, which can transmit a pathogen to a host).

If the disease, such as pneumonia, is constantly present in relatively low numbers, it is considered endemic. Often endemic diseases have reservoirs, or a site where the infectious agent can live without causing disease, and from which infection of others can occur. These reservoirs may be other animals (mice for the hanta virus, rats for typhus, cows for tuberculosis), or asymptomatic humans (*carriers* such as Typhoid Mary).

At other times, the *incidence* of a disease (the number of cases in a population) may suddenly increase. This would be considered an *outbreak* or, if the incidence became unusually high in a localized region, an *epidemic* of the given disease. An epidemic may be a *common-source epidemic,* in which a large number of individuals become ill from a single contaminated source (i.e., food poisoning following a company picnic). This will usually result in a very sudden increase, and later decrease, in the incidence of the infection. If the rise and subsequent fall in the number of infected individuals is slower, it is often due to a propagated, or *person-to-person epidemic* (i.e., influenza).

It is the role of the *epidemiologist* to determine the source(s) of an epidemic, how it has been spread, and how it can be controlled. Today's epidemiologist is not only interested in the incidence and spread of infectious disease, but also the cause of non-communicable diseases (cancer, heart disease, or the effect of environmental toxins) and injuries (extent of injuries when seat belts are used) in groups of people.

EPIDEMIOLOGICAL METHODS

To monitor the occurrence of disease, the epidemiologist relies heavily on statistical data. What is the *incidence rate* of the given disease? This involves determining the number of people who develop the disease in question during a certain period of time and dividing that value by the total number of people in the population. An increase in the incidence rate would indicate growth in the epidemic. It is also important to know the ***prevalence rate*** of the disease. This differs from the incidence rate in that it measures the exact number of people who have a given disease at a particular time. This value indicates the magnitude of the epidemic. The source of this data might be public records (vital statistics and census data), hospital records, questionnaires, or surveys.

Epidemiological studies may be descriptive, surveillance, field, or hospital studies. In ***descriptive studies,*** the location, time, individuals, procedures, and any common features are described. The morbidity (incidence of disease) and mortality (incidence of death) rates determined assist in providing a perspective on the increase or decrease of the disease and why the numbers are changing. Descriptive studies tend to be retrospective—looking back over what has already occurred. Their findings, however, are often able to be used to improve overall health care, taking into account not only the disease itself but also those environmental factors that contribute to its increased incidence.

Surveillance epidemiology tracks epidemic diseases and examines what changes are necessary to control its spread. These factors may include environmental change (improved water treatment methods), isolation of individuals who have been exposed, or vaccination. Surveillance methods were responsible for the eradication of smallpox and the control of malaria and polio.

When unexpected outbreaks of a disease occur, it often necessitates ***field epidemiology*** methods. Recent field studies include the identification of *E. coli* 01 57: H7 food poisoning and the Hanta virus outbreak in the southwest US. In both of these studies, it was the sudden increase in illness noted by hospitals that triggered the study. At other times, citizen complaints can be the initiation of a field epidemiologic study. An example would be the increased incidence of arthritis noted around Old Lyme, Connecticut which led to the identification of Lyme Disease.

The final type of epidemiological study is the ***hospital study.*** Nosocomial, or hospital-derived, infections are a major concern to health care facilities. The hospital environment is especially vulnerable to the appearance of infections for a number of reasons. Many patients have weakened resistance to infectious disease due to their illnesses. Patients, themselves, are often reservoirs of highly virulent pathogens. Cross-infection can readily occur because of the crowding of patients in rooms. Hospital personnel move between rooms and can serve as a vector of transmission. Many hospital procedures are invasive, overcoming the body's natural defense barriers. Often patients are receiving medication that is immunosuppressive, weakening their ability to fight off infections. Finally, the hospital environment contains many antibiotic-resistant organisms due to the frequent use of antibiotics. The result is that approximately 5% of all patients who are hospitalized will develop a nosocomial infection. To deal with these infections effectively requires the rapid identification of cases and the ***etiologic*** (causative) agent, the characterization of the epidemiologic features of the infection, the development and implementation of control methods, and constant monitoring to assure that control methods are being used effectively.

PUBLIC HEALTH AGENCIES

Overseeing the occurrence of illness and its control are a number of public health agencies and private health organizations. These include the World Health Organization, the Department of Health, Education and Welfare, the United States Public Health Service (the parent organization of the Center for Disease Control and Prevention or CDC), and state and local public health departments. The presence of national and world monitoring organizations facilitate the development of wide-range environmental and reservoireradication control methods. In our fast traveling society, we can no longer think of ourselves as isolated from those epidemics occurring on the other side of the world.

LABORATORY OBJECTIVES

In this exercise you will be simulating an epidemic amongst your laboratory group. Through this experience you should:

- Determine the source of a simulated epidemic.

- Understand how the science of epidemiology contributes to our understanding of disease.

- Understand how the epidemiologists collect information.

- Understand the types and uses of epidemiology—descriptive epidemiology, surveillance epidemiology, field epidemiology, and hospital epidemiology.

MATERIALS NEEDED FOR THIS LABORATORY

1. One Nutrient agar plate per person.

2. One latex glove or small plastic sandwich bag per person.

3. One numbered unknown swab per person. Select swabs will have been dipped into a broth culture of *Serratia marcescens*. The others have been dipped into sterile water which has had red food coloring added to it.

LABORATORY PROCEDURES

1. Obtain a nutrient agar plate and divide it into five sectors, numbering them from #1 to #5.

2. Obtain a numbered swab. Be sure to record your number!

3. Place the latex glove or sandwich bag on your left hand. This will enable the right hand to be used for swabbing. Swab the palm and fingers of your gloved hand with the swab. Discard the swab in a beaker of disinfectant.

4. Shake hands, using your gloved hand, with a classmate. Be sure that your fingers come in contact with the palm of the other person's gloved hand. After shaking hands, touch your fingers on the agar in the first sector on your petri plate. Be sure to record the swab number of the person you shook hands with.

5. Repeat step 4 with four other members of your class, being sure to record their swab number on the appropriate sector.

6. Discard your glove or sandwich bag in a Biohazard bag.

7. Incubate your plate at room temperature until the next laboratory period.

8. Following incubation, determine the presence of *Serratia marcescens* (red pigmented) colonies on your plates. Record the sector(s) in which they were located. Try to determine which swab(s) were contaminated and therefore the source of the "epidemic." The individual(s) with the original contaminated swab would be considered the index case(s) for the epidemic.

9. Record your results on the Laboratory Report Form.

NOTES

Name _____

Section _____

LABORATORY REPORT FORM

EXERCISE 27
EPIDEMIOLOGY

What was the purpose of this exercise?

Compare the following table with your results:

SECTOR NUMBER	SWAB NUMBER	PRESENCE OF *SERRATIA MARCESCENS*
#1		
#2		
#3		
#4		
#5		

Compare your results with other members of your class. Which swab(s) appear to have represented the index case of the epidemic?

What problems did you encounter in determining the index case of the epidemic?

What type of epidemiologic study (descriptive, surveillance, field or hospital) would this be considered?

QUESTIONS:

1. Name four portals of entry to the human body and give for each a disease-causing pathogen that would enter through that portal.

2. Often it is found that individuals who come in contact with an infected individual do not acquire the disease. What would be some explanations for this observation?

SEROLOGICAL REACTIONS

Serological reactions are specific reactions involving antigens and antibodies which occur *in vitro*, or in a test tube or other artificial environment. The term serological is used because serum is the most convenient source of antibody. Antibody-antigen reactions are typically very specific, often approaching the specificity of enzymatic reactions. It is because of this high level of specificity that serological reactions can be used for diagnostic work.

BACKGROUND

Antigens consist of a wide variety of molecules, including proteins, most polysaccharides, nucleoproteins, lipoproteins, and various small molecules attached to protein or polypeptide carrier molecules. These varied molecules have two properties in common. They are *immunogenic* (able to stimulate antibody formation) and exhibit specific reactivity with antibody molecules.

Antibodies are protein molecules produced by the body's immune system in reaction to exposure to a foreign or "non-self" antigen. They are characterized by containing at least two antigen binding sites. This ability to combine with two identical antigens forms the basis of many serological tests.

The interaction between antigen and antibody is usually highly specific. This means that if you know the identity of one component of an antigen-antibody reaction (either the antigen or the antibody), the other component can be detected with a very high level of sensitivity and reliability. If a reaction occurs when antibody is added to an unknown mixture, it is virtually certain that the antigen was present. When the term *homologous* is used to describe an antibody or an antigen, it indicates that the antigen and antibody can react with each other. That is, an antigen will react with its homologous antibody, and, barring cross reactions, only with its homologous antibody. Similarly, an antibody will react only with its homologous antigen.

There are many types of *in vitro* antigen-antibody reactions that can be used for the determination of the identity of a pathogenic organism or the presence of specific antibodies. These include agglutination reactions, precipitation reactions, fluorescent antibody reactions, neutralization reactions and complement-fixation reactions. They may be used to detect the presence of either antibody or antigen, depending upon the specific test being performed. For example, if you want to determine if a patient has been exposed to a certain virus (or bacteria, or rickettsia), you could mix the patient's serum with a known sample of antigen; if a reaction occurs, the antibody is present in your patient's serum. Conversely, an unknown virus can be tested against a known sample of antibodies; if reaction occurs, the identity of the virus will be known. The various types of *in vitro* serological reactions that are commonly used in diagnostic laboratories vary in their sensitivity and the ease with which they can be performed. In this exercise we will examine several applications of the *agglutination reaction*.

Agglutination occurs when a particulate antigen, such as cell wall components, flagella, or capsules, reacts with its homologous antibody and aggluti-

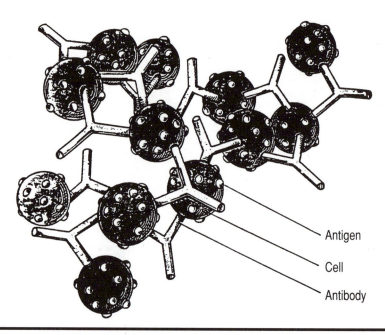

Figure 28.1   Agglutination Reaction

nates or clumps (See **Figure 28.1**). Characteristically, the antigen is *polyvalent* (containing many antigenic sites), while the antibody is in most cases bivalent (able to combine with two identical antigens).

When the two are mixed, increasingly large complexes of antigen-antibody are formed that appear to clump together. The agglutination complexes form because the bivalent antibody (Ab) molecule reacts with two particulate antigens (Ag). Each antigen, because it is polyvalent, can also react with several antibody molecules. The repeating complex eventually gets so large that the particles appear as visible clumps (see **Figure 28.2**).

Agglutination tests can be performed either on a slide or in a test tube. In a slide agglutination test, a drop of bacterial suspension might be combined with a drop of known antiserum for the rapid determination of homology. This can enable rapid preliminary identification of the pathogen. In other cases, a drop of the patient's serum can be combined with a known, and usually killed, suspension of bacteria to determine the presence of homologous antibodies in the patient. This usually indicates that the patient has been exposed to the specific pathogen, but does not necessarily indicate the source of a current infection.

To determine the exact level of antibody in serum, an agglutination titration must be performed. The antibody titer is an estimation of the concentration of antibody in the patient's serum. If an increase in titer is observed in successive tests, it can be assumed that the infection is caused by the test organism. The antibody titer rises relatively slowly in the early stages of an infection, rising to a maximum level and then falling off to a low level. Following a secondary exposure to the same antigen (an *anamnestic response*), the increase in antibody titer occurs much more rapidly and to a higher level.

Because of the rapidity and ease of performing agglutination reactions, passive agglutination methods have been developed. In passive reactions, noncellular antigens or antibodies are converted into a cellular or particulate form. It may involve the utilization of sheep red blood cells that have been treated to enable their absorption of specific antigens or the coating of latex beads with either antigen or antibody. These techniques have become the basis of most of the rapid clinical identification tests performed today.

Modern technology has also enabled the development of highly specific antibody molecules. Early testing for antigen identification used antibodies produced primarily by the introduction of antigens into an animal (for examples, goats) which would then produce the specific antibody. These antibodies could then be used in test systems. A definite drawback to this system was the possibly impurity of the antibody testing system. *Monoclonal antibodies* are used today for many of the antigenantibody test systems (see Figure 28.3). With monoclonal antibodies, the specific antigen is injected into a mouse which will produce antigen-specific antibodies. The antibody producing cells (plasma cells) in the mouse can then be harvested. These cells only produce antibodies against the specific antigen, however they are mortal cells. They can be combined with a myeloma cell (a cancer cell that multiplies rapidly and is immortal), forming a *hybridoma* or antibody-producing cell that undergoes frequent and repeated cell divisions. The antibodies produced by these cells are highly specific and provide increased sensitivity in testing systems.

SLIDE AGGLUTINATION REACTIONS— BLOOD TYPING

Human red blood cells contain varied glycoprotein and glycolipid molecules on their cell membrane surface. These molecules have specific antigenic properties. Karl Landsteiner was the first to note that if blood cells and serum from different individuals were combined, agglutination would occur. We now know that the basis of this agglutination is the structure of one specific glycolipid, known as substance H. which is found on the cell's surface. This glycolipid can be naked or combined with a mucopeptide (N-acetyl-galactosamine), or a sugar (galactose). Glycolipid plus N-acetylgalactosamine is commonly referred to as the A antigen. Therefore an individual who has this modification of substance H is said to have type A blood. If galactose has been added to the glycolipid, it is said to be the B antigen and the individual is said to have type B blood. Type AB blood is due to the presence of the N-acetylgalactosamine of some of the glycolipids, and galactose on others. Type O blood is characterized by the absence of either the A or the B antigen. These blood cells would, therefore, only contain the naked glycolipid.

It must be noted that the nature of substance H on red blood cells is only one of approximately 50 surface factors that have been identified as distinguishing individual red blood cells. Because it is one of the more prevalent antigenic forms, it has become the basis for our preliminary blood screening. The Rh_D antigen is also commonly screened for. When this antigen is present, the blood is said to be Rh+; if absent, it would be Rh-.

A patient's blood type may be determined by mixing the patient's (the unknown) red blood cells with serum known to agglutinate either Group A, Group B or Group D (Rh+) cells. The term *hemagglutination* is commonly applied to those agglutination reactions which involve red blood cells. Blood typing is almost always done when a person enters a hospital, especially if there is a possibility of surgery, or when there may be questions about the blood type of a newborn infant.

Agglutination With	Indicates
Anti- A	Type A
Anti- B	Type B
Anti- A and Anti- B	Type AB
Neither Anti- A nor Anti- B	Type O
Anti- D (Anti-Rh)	Type D (Rh+)

Failure to agglutinate with the Rh antiserum indicates Rh - (type D)

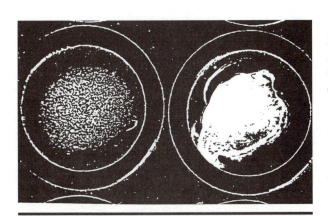

Figure 28.2   Positive (left) and Negative (right) Agglutination Reactions

SLIDE AGGLUTINATION REACTIONS— SEROLOGICAL IDENTIFICATION OF BACTERIAL UNKNOWNS

The traditional identification of bacterial organisms involves their overnight cultivation, followed by the performance of specific biochemical tests and staining techniques. Whereas this method tends to be quite accurate, it is time consuming and expensive. *Serotyping,* or the determination of antigens for the purpose of identification, is a much faster and easier method. The patient's serum can be tested against commercially available antigens to determine the

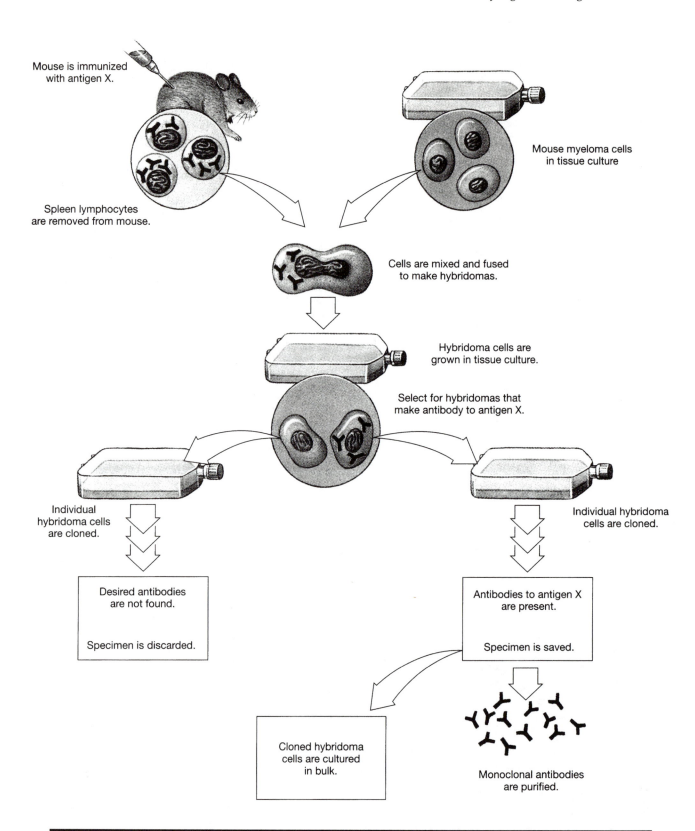

Figure 28.3   Production of Monoclonal Antibodies

presence of specific antibodies; pathogens can be cultivated and isolated and then tested using commercial antisera to determine their specific identity; or a specimen from the patient can be directly tested for the presence of specific antigens. An additional benefit to serotyping can be seen when tracing the course of an epidemic. For example, over a thousand distinctive antigenic variations of *Salmonella* have been identified. By determining the exact serovar (antigenic type) of *Salmonella* involved in an infection, it becomes possible to determine relatedness of cases in the outbreak.

Bacterial cells have two major types of antigens on their cell surface. Both types are species, and sometimes strain, specific. These antigens are found either on the outer membrane of gram-negative cells or associated with the flagella. The antigens associated with the outer membrane are commonly referred to as somatic or "O" antigens. The designation O comes from the German *ohne hauch* or 'without spreading.' This is to contrast it with the flagellar or "H" antigen, which is named from the German *hauch* (spreading).

Somatic "O" antigens are heat stable polysacchrides. Each form known has been given a number from 1 to 64. It is possible to group organisms in serogroups based upon the presence of common antigens. Each member of a serogroup must have at least one antigen in common. For example, all members of Serogroup A have antigen 2.

The Flagellar or "H" antigens are heat labile proteins associated with the flagella. They are commonly designated by lower case letters (i.e., group d).

When a laboratory wishes to serotype an unknown, the technician will usually determine the serotypes of both the O and H antigens present on the bacterial cell surface.

In this exercise we will use a polyvalent antibody that reacts with almost all strains of *Salmonella*. This polyvalent antibody is really a mixture of O antibodies and is used to confirm that the bacterial isolate actually is *Salmonella*. If, in the clinical laboratory, a positive reaction occurs with the polyvalent antibody, then the unknown is tested with group specific O antibodies. Once the O antigens have been identified by slide agglutination, then the identity can be confirmed by the identification of the H antigens. This is a tube agglutination test.

FEBRILE AGGLUTINATION TEST

The term febrile antigens are generally accepted as referring to bacterial suspensions representative of a number of microbial species pathogenic to man and characterized by causing a fever in their host. The most distinctive organisms in this group are members of the genera *Salmonella, Brucella, Francisella, Leptospira,* and some members of the *Rickettsia.*

The febrile agglutination test uses known antigens to determine if a patient's serum contains antibody to one or more of the antigens contained in the set. The antigens are bacterial cells that have been killed and stained. In this test, the antigen is known and you will test for the presence of homologous antibody in an unknown (a patient's serum). Known positive controls should always be run with the test. A typical febrile agglutination test set (i.e., Difco #2407-32-7) includes the following cellular antigens and controls:

Proteus OX19 antigen
Brucella abortus antigen
Salmonella 0, Group D (Typhoid O) antigen
Salmonella H Group a (Paratyphoid A) antigen
Salmonella H Group b (Paratyphoid B) antigen
Salmonella H Group d (Typhoid H) antigen
Positive control serum
Negative control serum

Agglutination with any of the antigens in the set indicates that the patient has been exposed to the organisms and may recently have had, or now has, the indicated disease.

This test is usually run by mixing one drop of the patient's serum with one drop of each of the test antigens. If agglutination occurs, and all the controls show the expected reactions, the patient is presumed to have been exposed to the antigen. (How else could the patient have the antibody?) Additional testing is usually necessary to determine if the patient is harboring the infectious agent and to obtain more specific information about which strain of antigen the patient was exposed to.

One of the most commonly used agglutination reactions is the typing (A, B. O. and Rh + or D) of blood. In this exercise you will determine your blood type. You will also use agglutination reactions to serotype bacterial unknowns. (What does serotype mean?) The febrile agglutination test will be used to demonstrate how known antigen (killed bacterial cells) can be used to detect antibodies in a patient's serum. In all types of serological reactions, however, the antibody and antigen must react. These reactions are highly specific, and it is this specificity that makes the reactions so important diagnostically.

OBJECTIVES

Serological reactions are characteristically very specific and, therefore, of diagnostic value. In this exercise, you should:

- Understand the specificity of the antigen-antibody reaction.

- Discuss the antigenic basis of blood typing.

- Know how to read and interpret an agglutination reaction.

- Appreciate the diagnostic value of serological reactions.

- Understand why a known antigen or antibody can be used to identify an unknown sample of virus or serum.

MATERIALS NEEDED FOR THIS LABORATORY

A. BLOOD TYPING

1. Anti-A, Anti-B, and anti-Rh typing sera.

2. Glass slides.

3. Lancets and alcohol (70%) swabs or wipes.

4. Toothpicks or applicator sticks.

5. **Optional: artificial or purchased sera may be used for blood typing in place of student blood.**

B. SEROTYPING OF UNKNOWN BACTERIA

1. Polyvalent *Salmonella* O Antiserum Set A-1 (Difco No. 2892-32: *Salmonella* Poly A- 1) and Positive Control Antigen (Difco No. 2840-56: *Salmonella* O Antigen, Group B).

2. Phenolized saline suspensions of *Salmonella*. If you are instructed to prepare your own suspension, you will need a 24 hour agar slant culture of the organism and at least 1.0 ml of phenolized saline (0.5% phenol in 0.85% saline).

3. Glass slides or depression slides

C. FEBRILE AGGLUTINATION TEST

1. Febrile agglutination sets (i.e., Difco No. 2407-32-7), including positive and negative controls.

2. Unknown serum samples (may be obtained from local hospitals or clinics)

3. Agglutination plates or slides

PROCEDURES BLOOD TYPING

1. Obtain a glass slide and divide it into three sections with a wax pencil. See **Figure 28.4.**

2. Obtain samples of the three antisera (anti-A, anti-B, and anti-D). Have them ready to use.

3. Obtain a lancet. Open it and position it so that the tip does not touch the laboratory bench. (Leave the point in the package until you are ready to use it.)

4. Carefully clean the tip of your finger with the alcohol swab or wipe. Allow the cleansed area to air dry. (Do not recontaminate it by drying it with a paper towel or cloth.)

5. Lance the fingertip by quickly stabbing it with the lancet. A quick jab is usually least painful and most effective.

6. Allow one drop of blood to fall into each of the circles on the slide. Press a sterile swab against the lancet wound until a clot forms and blood flow stops (about one minute). Dispose of the lancet and swabs in a biohazard container, as directed by your instructor.

7. Drop one drop of the antisera into the appropriate circles from about 1 cm above the slide. Do not allow the dropper to touch the blood.

8. Using separate toothpicks for each of the three mixtures, carefully mix the blood and antiserum to obtain a homogeneous mixture. Agglutination should occur (if it is going to) almost immediately.

9. Record your results on the Laboratory Report Form, Part A.

B. SEROLOGICAL IDENTIFICATION OF BACTERIAL UNKNOWNS

1. If you are not supplied with a suspension of bacteria in phenolized saline, you may prepare

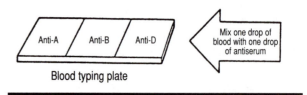

Figure 28.4   Slide-Agglutination Slide

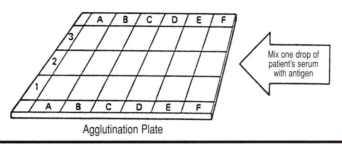

Figure 28.5   Slide-Agglutination Plate

your own suspensions. Place about 1 ml of the phenolized saline into a clean test tube. Transfer a loopful of the unknown bacterial culture (from an agar slant) to the phenolized saline and mix well.

2. Divide a clean glass slide into three sections with a wax pencil.

3. Each of the sections will contain different combinations of test and control mixtures. Usually one drop is a sufficient amount of reagent. As before, drop the test and control reagents onto the slide—do not allow droppers to touch the slide or the reagents.

4. Prepare the test mixtures according to the table:

5. Mix the reagents in each of the circles with separate toothpicks. Agglutination should occur almost immediately.

6. The first mixture, with saline and the unknown bacterial suspension, should not agglutinate (negative control). The third mixture, with the antiserum and the suspension of known positive bacteria, should agglutinate (positive control). The second mixture will agglutinate only if the organism has antigens that are homologous to one of the antibodies in the polyvalent mixture on its flagella or outer membrane.

7. Record your results on the Laboratory Report Form, Part B.

C.   FEBRILE AGGLUTINATION TEST

1. There are six bacterial antigens in the set, and each one of them must be tested against the unknown serum and against the two controls.

2. Depending upon the type of plates or slides available, mark three rows of six test areas. One row will be used for the unknown serum and the other two will be used for the controls. See **Figure 28.5**

3. Place one drop of serum into each test area in the appropriate row.

4. As before, drop one drop of antigen into the drop of serum. Do not allow the dropper tip to touch the plate or the drop of serum.

5. Mix each with a separate toothpick or applicator stick.

6. Agglutination should occur almost immediately; the results should be observed in about one minute.

7. Record your results on the Laboratory Report Form, Part C.

NOTES

LABORATORY REPORT FORM

EXERCISE 28
SEROLOGICAL REACTIONS

What was the purpose of this exercise?

A. BLOOD TYPING

1. Complete the following table:

ANTISERUM	HEMAGGLUTINATION
Anti-A	
Anti-B	
Anti-D	

Based on the results you got, what is your blood type?

2. Determine the blood type distribution for your class. Obtain the data from the other students in the class:

NUMBER OF STUDENTS IN CLASS: _____

Blood Type	Number	Percent
Type A		
Type B		
Type AB		
Type O		
Type Rh+ (D)		
Type Rh- (d)		

B. SEROLOGICAL IDENTIFICATION OF BACTERIAL UNKNOWNS

1. Describe the agglutination reactions observed in the slide agglutination serotyping of the bacterial cultures.

2. If this were a clinical unknown, would additional serological testing be necessary?

C. FEBRILE AGGLUTINATION TEST

1. Fill in the following chart using your data from the febrile agglutination test:

ANTIGEN	REACTION (+ OR -)
Proteus OX-19	
Brucella abortus	
Salmonella O (D)	
Salmonella H (a)	
Salmonella H (b)	
Salmonella H (d)	

QUESTIONS:

1. What type of antibodies might be found in the serum of a person with type B blood cells?

2. Why are agglutination reactions useful for the identification of microorganisms?

3. Distinguish between O and H antigens.

4. How could you modify a slide agglutination test to determine the titer of antibody present when using a known antigen?

5. Seven individuals have been found to have a food-borne *Salmonella* infection which has been traced to a common source. Once this information became publicized, several additional people went into local emergency rooms, complaining of similar symptoms. How could you determine if the source of their infection is likely the same as that of the original seven individuals?

THE COLLECTION AND TRANSPORT OF CLINICAL SPECIMEN

BACKGROUND

In order to successfully determine the causative agent of a bacterial infection, clinical specimen must be properly collected, transported and examined. Specific methods will vary dependent upon the site being sampled, but in all cases it is important that aseptic techniques be used to prevent contamination with normal flora. Common sources of contamination are shown in Table 29.1.

Table 29.1

Source of Culture	Possible Contamination Sources
Urine	Urethra and perineum
Blood	Skin at site of venipuncture
Middle Ear	Outer ear canal
Nasal sinus	Nasopharynx
Throat	Oral cavity and gingiva
Subcutaneous wound or abscess	Skin and mucous membranes

A. Specimen

The types of specimen that can be examined include those which can be directly sent to the laboratory (urine, blood, cerebral spinal fluid, fecal samples, sputum, skin scrapings), as well as those that must be swabbed (abscess, ear, genital, nasopharynx, throat). The desired sample is collected in a sterile container or with a sterile swab, avoiding contact with any surrounding tissue. Once the specimen is collected, it is important that it be promptly sent to the lab and cultured.

For these specimen to be accepted and examined by the lab, several criteria must be met. It must be properly labeled, promptly transported, properly sealed, and contained within a suitable transport system for the type of specimen collected.

B. Transport

To help preserve the quality of the swab specimen, preventing the death of fastidious organisms and the overgrowth of normal flora, it is helpful to utilize one of the several transport systems available. These transport systems include both aerobic and anaerobic transportation and the incorporation of one of several transport mediums. *Transport media* are specially designed to preserve organisms during transport. They consist of buffered salts that prevent microbial growth, water to prevent dehydration, and oftentimes reducing agents to support the survival of anaerobic organisms. These transport systems are commercially available under the trade names of Culturette, Anaerobic Culturette, Transwab, Transtube, CultureSwab, Port-A-Cul System, as well as several others. We will utilize the Culturette and Anaerobic Culturette systems for this exercise.

Specimen suspected of containing *Shigella* species must be processed immediately. If this is not possible, the transport system should be refrigerated until culturing is possible. If *Neisseria gonorrhoeae*, *Neisseria meningitidis*, or *Haemophilus influenzas* are suspected, the samples must not be refrigerated, as these organisms are highly susceptible to cold.

C. Examination

Once the specimen is received by the laboratory, it will be microscopically examined and cultured on suitable media as determined by the source of the inoculum and the suspected infectious agent. Processing of specimen for bacterial agents will include the utilization of Gram staining, enrichment broths, blood agar, chocolate agar, MacConkey agar, EMB agar, phenylethyl alcohol agar, Hektoen enteric agar, or Salmonella-Shigella agar. What type of infectious agent would each of the above media be used to detect?

We will be utilizing a simulated wound to examine the effects of transport on successful recovery of pathogens from cultures.

OBJECTIVES

- Understand the importance of proper collection and transport of microbial specimen to the laboratory.

- Evaluate the factors during transport that might alter the results of microbial testing.

- Evaluate the usage of aerobic and anaerobic transport systems and transport media.

MATERIALS NEEDED FOR THIS LABORATORY

1. Twenty-four-hour cultures of:
    Clostridium sporogenes (thioglycollate broth)
    Staphylococcus epidermidis (Tryptic soy broth)

2. Sterile test tubes (2)

3. Sterile Pasteur pipet and bulb (2)

4. Tryptic Soy Agar plates (6)

5. Calcium alginate swabs (2)

6. Culturette (Becton Dickinson)

7. Anaerobic Culturette (Becton Dickinson) (1)

8. GasPak anaerobic jar, with catalyst, indicator, and gas generator envelope

LABORATORY PROCEDURE

1. Prepare mixed culture of *Clostridium sporogenes* and *Staphylococcus epidermidis* by trans-
fering 5 – 10 drops of each culture to a sterile test tube. Use a separate sterile Pasteur pipet for each transfer.

2. Using a sterile calcium alginate swab, dampen the swab in the mixed culture. Ring the inside of the tube with the swab to avoid excess moisture from dripping on your plates. Inoculate two Tryptic Soy Agar plates, utilizing the swab for the first section and your sterile inoculating loop for the second and third sections as you streak for isolation.

3. Incubate one labeled plate at 37°C under aerobic conditions. Place the second labeled plate in the GasPak and incubate it anaerobically at 37°C until the next laboratory period.

4. Moisten a second swab in the broth culture. This swab is to be placed in a sterile test tube and incubated at room temperature until the next laboratory period.

5. Following the directions for the Culturette system, moisten the swab in the broth culture. Replace the swab in the transport tube, break the ampule, label, and store at room temperature until the next laboratory period.

6. Following the directions for the Anaerobic Culturette system, moisten the swab in the broth culture. Replace the swab in the transport tube, break the transport medium and anaerobic generating system ampules, label, and store at room temperature until the next laboratory period.

7. Examine the TSA plates incubated aerobically an anaerobically. Perform Gram stains on isolated colonies to determine identity of organism(s). Record you results on the Result page. Save the plates in the refrigerator until the next laboratory period.

8. Using alcohol sterilized forceps, remove the swab from the sterile test tube stored at room temperature, inoculate two Tryptic Soy Agar plates, utilizing the swab for the first section and your sterile inoculating loop for the second and third sections as you streak for isolation. Incubate one labeled plate at 37°C under aerobic conditions. Place the second labeled plate in the GasPak and incubate it anaerobically at 37°C until the next laboratory period.

9. Using the Culturette swab, inoculate one Tryptic Soy Agar plate, utilizing the swab for the first section and your sterile inoculating loop for the second and third sections as you streak for isolation. Incubate the labeled plate at 37°C until the next laboratory period.

10. Using the Anaerobic Culturette swab, inoculate one Tryptic Soy Agar plate, utilizing the swab for the first sections and your sterile inoculating loop for the second and third sections as you streak for isolation. Incubate the labeled plate in the GasPak system at 37°C until the next laboratory period.

11. Examine the swab, Culturette, and Anaerobic Culturette TSA plates incubated aerobically and anaerobically. Perform Gram stains on isolated colonies to determine identity of organism(s). Compare these plates with the plates from day one which have been stored in the refrigerator. Record you results on the Laboratory Report Form.

NOTES

Name _____

Section _____

LABORATORY REPORT FORM

EXERCISE 29
PHYSICAL CONTROL METHODS

What was the purpose of this exercise?

1. Complete the following table indicating the presence (+) or absence (–) of growth.

	Initial Aerobic Incubation	Initial Anaerobic Incubation	Swab- Aerobic Incubation	Swab- Anaerobic Incubation	Aerobic Culturette Incubation	Anaerobic Culturette Incubation
Clostridium sporogenes						
Staphy. epidermidis						

What effect did the varied storage methods have on the maintenance of organisms from collection to culturing?

QUESTIONS

1. Give three reasons why a specimen should be transported to the laboratory and processed as quickly as possible.

   a.

   b.

   c.

2. What type of media would a properly transported sample be plated on for the determination of the presence of the following species?

   a. *Escherichia coli*

   b. *Staphylococcus aureus*

   c. *Salmonella* sp.

3. How important is the utilization of anaerobic incubation of wound samples?

EXERCISE 30

THROAT CULTURES

CAUTION: The throat may contain a variety of both pathogenic and opportunistic organisms. Be very careful in all inoculations. Handle all cultures, both those from the throat cultures and any demonstration plates, with care.

BACKGROUND

The upper respiratory tract has a rich population of indigenous organisms, most of which are considered to be normal flora. It also supports the growth of several important pathogens, the most important of which include the streptococci, the staphylococci, the neisseriae, the diphtheroids, yeasts (particularly *Candida*), and enteric gram-negative rods. In addition to these, which are commonly isolated from the upper respiratory tract of persons of all ages, certain other pathogens are occasionally encountered in samples from the very young (under five years) and the elderly (over sixty years). These latter include the genera *Hemophilus, Bordetella, Yersinia,* and *Francisella.*

THROAT CULTURES: SITE TO BE SAMPLED

Throat cultures are taken from the area at the very rear of the mouth, behind the uvula by rotating a sterile swab over the area (See **Figure 30.1**). Any inflamed area that might be observed should be cultured. It is very important that:

1. The culture be taken from the correct part of the oral cavity. The single most common error

encountered in throat cultures is that the culture is taken from the wrong part of the mouth. The swab should be rotated over the mucosal surface behind the uvula.

2. The swab must be rotated both when the sample is taken and when the swab is used to inoculate the media. This will ensure that a large enough surface area will be sampled and that the media will be adequately inoculated.

TECHNIQUES FOR MEDIA INOCULATIONS

In general, techniques for the isolation and identification of the pathogenic forms have been well

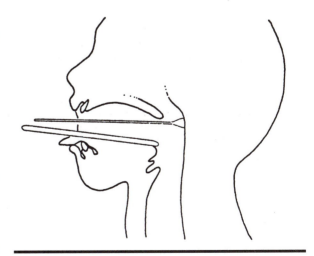

Figure 30.1   Procedure for Taking a Throat Culture

245

defined, and the procedures are straight-forward and direct. Much of the diagnostic work involves distinguishing between the pathogenic and nonpathogenic species that comprise the normal flora. The use of special differential and selective media has greatly facilitated this task. The correct method for inoculating diagnostic media when the sample is contained on a swab is to rotate the swab on the surface of the medium so that the entire surface of the swab has had contact with the medium. You should cover an area about the size of a nickel or quarter and then use an inoculating loop to complete a steak plate (See Exercise 8).

The following selective and differential media are commonly used for routine throat cultures:

1. Blood agar

2. Chocolate agar

3. Mueller tellurite agar

4. Mannitol-salt agar

SELECTIVE AND DIFFERENTIAL MEDIA AND THEIR REACTIONS

Several of these media will be used in this exercise. You will be asked to take a throat culture and to compare the growth obtained from it with the characteristic colonies produced by known and representative organisms. The selective and differential media you will use in this exercise include:

1. **Blood Agar.** Blood agar is a nutrient medium, such as trypticase soy agar, that has whole sheep red blood cells added to it. The most common mixture uses a 5% suspension of specially washed cells in the medium. Of course, special preparatory procedures must be followed to prevent damage to the red blood cells by the heat of the melted medium. It is not uncommon to add special nutrients and/or selective agents to blood agar, depending upon the needs of the laboratory.

Bacteria that secrete hemolysins (hemolytic enzymes) are able to lyse the sheep red blood cells in the medium. The destruction of these cells will produce distinct zones of hemolysis around their colonies. They may be categorized according to the type of hemolysis they cause:

- Alpha-hemolysis: an incomplete lysis of the erythrocyte and a partial denaturation of the hemoglobin to various heme products. Alpha-

hemolysis results in green discoloration of the area around the colony. Microscopic examination would reveal cellular debris in the green-colored regions.

- Beta-hemolysis: a complete lysis of the erythrocyte membrane and the reduction of hemoglobin to colorless products through the release of hemolysin enzyme from the bacterial cell. Beta-hemolysis results in a completely cleared zone around the colony, and no cellular debris would be observed upon microscopic examination.

- Nonhemolytic: no visible change in the erythrocytes around the colonies. Sometimes nonhemolytic streptococci are referred to as *gamma-hemolytic,* but since there is no change in the erythrocytes, *nonhemolytic* is a more accurate description.

2. **Chocolate agar.** Chocolate agar is blood agar that has been heated to denature the proteins of the sheep red blood cells. This causes them to lyse and to turn a light brown color, not unlike that of milk chocolate (hence the name). Most formulas for chocolate agar include certain enrichment factors (such as added hemoglobin and other factors) that encourage the growth of some of the more fastidious pathogens, including *Neisseria* and *Haemophilus.*

Some bacterial hemolysins that are able to denature hemoglobin in blood agar are also able to cause changes in the color of chocolate agar. Many streptococci and lactobacilli that cause either alpha- or beta-hemolysis will cause either a greening or a decoloration of the light brown color in chocolate agar. However, if the bacteria being isolated are nonhemolytic, there will be no color change in the medium, and the colonial morphology will be typical of whatever species is being isolated.

3. **Mueller tellurite agar.** Mueller tellurite agar is a selective and differential medium for the identification of *Corynebacterium diphtheriae.* This medium inhibits the growth of most cocci, non-diphtheroid bacilli and yeast and allows the differentiation of the mitis, gravis and intermedius types of *C. diphtheriae.* The differentiation is based on the size and degree of darkening of the colonies.

4. **Mannitol-salt agar.** Mannitol-salt agar is a medium containing the sugar mannitol and 7.5% sodium chloride (salt). The salt inhibits the growth of most bacteria except the staphylococci. The mannitol dif-

ferentiates between *S. aureus* (a mannitol-fermenter) and *S. epidermidis* (a mannitol-nonfermenter). Organisms that ferment mannitol will produce a yellow zone around the colony as a result of the production of acids during fermentation. Organisms that do not ferment mannitol will produce a deeper red color (alkaline reaction) or will not change the color at all.

DIFFERENTIATION AND IDENTIFICATION: SCREENING AND CONFIRMING TESTS

1. **Streptococci from other gram-positive cocci.** All members of the genus *Streptococcus* are catalase negative, and all other gram-positive cocci are catalase positive. The catalase test is performed by either placing one drop of hydrogen peroxide on a colony or by picking up an isolated colony from a plate and placing it in a drop of hydrogen peroxide on a microscope slide (See Exercise 14). Remember, you do not want to perform the catalase test on colonies on blood agar (Why?). Catalase produced by positive organisms will cause the hydrogen peroxide to be broken down to water and oxygen, resulting in vigorous bubbling. Catalase-negative organisms do not cause bubbling.

2. **Staphylococci from micrococci.** All of the staphylococci are facultative, while most of the common micrococci are obligate aerobes. The two genera can be differentiated by determining their growth characteristics in thioglycollate broth (See Exercise 9). The staphylococci will grow throughout the media, while the micrococci will only grow at the top. In addition, the staphylococci are usually oxidase negative whereas the micrococci are usually oxidase positive.

3. *Staphylococcus aureus* **from** *Staphylococcus epidermidis*. Any one of the following tests can be used to differentiate between *S. aureus* and *S. epidermidis*. The coagulase test, described in Exercise 14, is considered the most important and is the one most frequently used.

Test	*S. aureus*	*S. epidermidis*
Coagulase	+	-
Acid from mannitol	+	-
Liquifies gelatin	+	-
DNase positive	+	-

4. **Alpha-hemolytic streptococci.** Many of the alpha-hemolytic streptococci are part of the normal flora. However, many of them can be pathogenic when they are introduced into other parts of the body. One of them, *Streptococcus pneumoniae*, is the causative agent of bacterial pneumonia and can be a very serious pathogen. It must always be ruled out when alpha-hemolytic streptococci comprise a significant proportion of the isolates from a throat culture. Virtually all the alpha-streptococci except *Streptococcus pneumoniae* are resistant to lysis by bile-salt solutions and will grow in the presence of such salts. This characteristic is used to determine if any of the alpha-hemolytic streptococci that are cultured from the throat are *Streptococcus pneumoniae*. The sensitivity of the bacterium to bile salts can be measured by either placing one drop of bile-salt solution over a colony or by using prepared disks (Taxo P or Optochin). Sensitive organisms will dissolve in the bile solution or will not grow adjacent to the disk. The characteristic gram-positive diplococcus morphology of *Streptococcus pneumoniae* should always be confirmed. (How?)

5. **Beta-hemolytic streptococci.** The beta-hemolytic streptococci are *all* pathogenic and must be carefully identified. One of the most important differential tests is the sensitivity of group A streptococci to the antibiotic Bacitracin. This sensitivity is measured by placing a disk with the antibiotic on blood agar that has been streaked with the beta-hemolytic isolate. Further differentiation of the other groups (B. C, and D) of beta-hemolytic streptococci requires the determination of several biochemical characteristics.

Because of the need for accurate and often rapid identification of the group A streptococci, several immunological tests are commonly used in the clinical laboratory. They do not require the growth of the organism and most can be performed directly on the throat culture swab. The group A antigen is extracted from any cells present and its presence is detected through agglutination with reactive latex (latex molecules which contain the specific group A antibody).

6. **Enterococci.** Many, but not all, of the enterococci are nonhemolytic. These organisms are part of the normal flora of the intestinal tract, and like the alpha-hemolytic streptococci from the throat, can be important pathogens when introduced into other parts of the body. The enterococci can be distinguished from all of the other streptococci, regardless of hemolytic characteristics, by their growth in 6.5%

salt, and the hydrolysis of esculin. Since the entero-cocci may exhibit alpha- and beta-hemolysis, the presence of group D enterococci must be considered. These three tests are usually considered sufficient to confirm the presence of one of the enterococci, regardless of the type of hemolysis observed.

7. **Neisseriae.** Since up to half the colonies that appear from a throat culture can be members of the genus *Neisseria,* selective media must be used to isolate the two pathogenic species: *Neisseria gonorrhoeae* and *Neisseria meningitidis.* Growth on either of these media, when confirmed as being oxidase-positive, gram-negative diplococci, is considered presumptive for either of the species. Also, neither of them will grow on ordinary nutrient agar, while most of the nonpathogenic forms will. These growth characteristics are used as part of the diagnostic protocol for the neisseriae. Also included with the neisseriae is *Moraxella (Branhamella) catarrhalis.*

8. **Diphtheroids.** The diphtheroids, like the neisseriae, are among the most prominent isolates from healthy throats. As with the neisseriae, identification of the pathogen *Corynebacterium diphtheriae* requires a special isolation medium and additional biochemical tests. Recognition of the typical gram-positive rod cellular morphology of the diphtheroids and the formation of metachromatic granules is critical for the correct identification of these organisms.

Table 30.1 provides a dichotomous key to assist in the differentiation of the organisms commonly encountered in throat cultures.

OBJECTIVES

In this exercise you will obtain a throat culture and compare the growth obtained with known cultures provided to you. This exercise will require that you:

- Take a throat culture.

- Examine the growth that develops on the media used and attempt to recognize colonial characteristics typical of those media.

- Acquire an understanding of the proper protocol for the identification of the organisms typically encountered in normal and abnormal throat cultures.

- Acquire an understanding of the nature of the normal flora of the throat and how it is distinguished from the pathogenic forms that might colonize that area.

- Develop an understanding of the routine diagnostic tests used to characterize the organisms used in this exercise.

MATERIALS NEEDED FOR THIS LABORATORY

1. Media:
   Blood Agar Plate
   Chocolate Agar Plate*
   Mueller Tellurite Agar Plate*
   Mannitol-Salt Agar Plate
   *NOTE: The Chocolate and Mueller Tellurite

Table 30.1   Dichotomous Key—Bacteria Commonly Encountered in Throat Cultures

I. Gram-negative	
A. Grows on Chocolate agar, but not on nutrient agar	
1. Ferments glucose and maltose.	*N. meningitidis*
2. Ferments glucose, but not maltose	*N. gonorrhleae*
B. Grows on nutrient agar, but not on Thayer-Martin agar	
1. Ferments at least one sugar	*Neisseria* sp.
2. Does not ferment any sugar.	*M. catarrhalis*
II. Gram-positive	
A. Catalase positive	
1. Aerobic	*Micrococcus sp.* (Not *Staphylococcus*)
2. Facultative	
a. Coagulase positive, ferments mannitol	*S. aureus*
b. Coagulase negative, does not ferment mannitol.	*S. epidermidis*
B. Catalase negative	
1. Grows in 6.5% salt and in medium containing bile	*Enterococcus*
2. Alpha-hemolytic	
a. Soluble in bile solution	*S. pneumoniae*
b. Not soluble in bile solution.	*Streptococcus* (Not *S. pneumoniae*)
3. Beta-hemolytic	
a. Bacitracin sensitive	*Streptococcus* (Group A)
b. Bacitracin resistant	*Streptococcus* (Group B. C, or D)

Agar may be prepared as a biplate.

2. Sterile calcium alginate swabs.

3. Known cultures:
Representative cultures, including some of the following will be available either as broth cultures for your inoculation onto the designated media, or for observation of their growth on the differential media.
Moraxella (Branhamella) catarrhalis
Corynebacterium diphtheriae
Neisseria gonorrhoeae
Neisseria meningitidis
Staphylococcus aureus
Staphylococcus epidermidis
Streptococcus faecalis
Streptococcus pneumoniae
Streptococcus pyogenes
Alpha-hemolytic Streptococcus

4. Optochin differentiation disks

5. Bacitracin differentiation disks

6. Reagents to complete testing:
Hydrogen peroxide
Coagulase plasma
Oxidase reagent

LABORATORY PROCEDURE

1. Obtain the following media
   a. One of each of the following for each throat culture:
   Blood agar plate
   Chocolate agar
   Mueller tellurite agar
   Mannitol-salt agar
   b. One Blood agar plate for each known streptococci assigned
   c. One Chocolate agar plate for each known neisseriae assigned
   d. One Mueller tellurite agar for each known diphtheroid assigned
   e. One Mannitol-salt agar for each known staphylococci assigned

2. Prepare streak plates of assigned known organisms on the appropriate agar.

NOTE: Your laboratory instructor may provide you with demonstration plates of the known organisms as many of them are pathogenic.

3. Using a sterile calcium alginate swab, culture the nasopharyngeal area. Be sure to rotate the swab in the tonsilar area, carefully avoiding the uvula. (See Figure 30.1)

4. Referring to **Figure 30.2**, prepare streak plates on each of the designated media. Use the throat swab for the primary inoculation and your sterilized inoculating loop for isolation. Dispose of the swab as indicated by your instructor.

5. On the Blood agar plate, place an Optochin differentiation disk on one side of your primary streak and a Bacitracin differentiation disk on the other end of the primary streak (See Figure 30.2).

6. Incubate all plates at 37°C for 24 – 48 hours.

7. Examine all plates and record the colony morphology observed. Be sure to note any reaction to the Optochin and Bacitracin disks.

8. Prepare gram stains of representative colonies.

9. Perform those additional tests needed to aid in the identification of your isolates. Refer to the dichotomous key in Table 30.1 to assist you.

10. To perform the catalase test: Using you sterile loop, place some cells from an isolated colony on a clean glass slide. Add a few drops of hydrogen peroxide to the cells. If bubbles form, the organism is catalase positive. Place the slide in a beaker of disinfectant to kill the organisms.

11. To perform the oxidase test: Place a strip of filter paper on a glass slide. Moisten it with oxidase reagent. Using your sterile loop, rub some cells from your isolated colony on the filter paper. Observe for the production of a dark pink to purple color indicating that the organism is oxidase positive. Dispose of the filter paper as indicated by your instructor.

12. Record all results on the Laboratory Report Form.

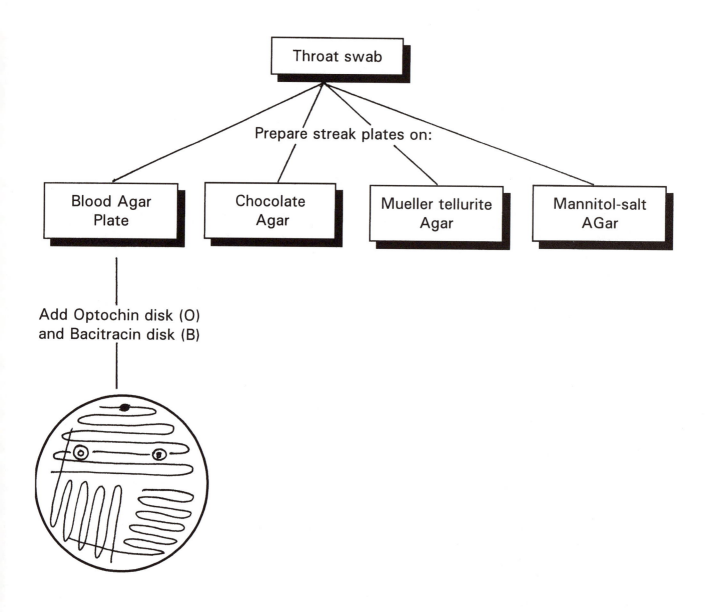

Figure 30.2   Throat Culture Protocol

LABORATORY REPORT FORM

EXERCISE 30
THROAT CULTURES

What was the purpose of this exercise?

Complete the following for those known organisms observed:

ORGANISM	MORPHOLOGY	GRAM REACTION	MEDIA	COLONY DESCRIPTION	OTHER TEST RESULTS

Throat Culture results:

Blood Agar

COLONY DESCRIPTION	MICROSCOPIC OBSERVATION	HEMOLYSIS TYPE	OPTOCHIN TEST	BACITRACIN TEST	CATALASE REACTION	OXIDASE REACTION

Identification of organism(s):

Chocolate Agar

COLONY DESCRIPTION	MICROSCOPIC OBSERVATION	ADDITIONAL TEST RESULTS

Identification of organism(s):

Mueller Tellurite Agar

COLONY DESCRIPTION	MICROSCOPIC OBSERVATION	ADDITIONAL TEST RESULTS

Identification of organism(s):

Mannitol-Salt Agar

COLONY DESCRIPTION	MICROSCOPIC OBSERVATION	ADDITIONAL TEST RESULTS

Identification of organism(s):

QUESTIONS

1. Distinguish between alpha- and beta-hemolysis.

2. How would you differentiate between *Staphylococcus aureus* and *Streptococcus pneumoniac*?

3. Why is it important to be able to rapidly differentiate between *Staphylococcus aureus* and *Staphylococcus epidermidis?*

4. You observed the following results on a throat culture isolate:

Gram-positive diplo- and streptococci, alpha hemolytic, Optochin positive, catalase negative

What is the identity of the organism?

EXERCISE 31

URINARY TRACT CULTURES

BACKGROUND

Urine from a healthy person, when carefully taken as a clean midstream or catheterized sample, will be found to contain remarkably few bacteria. One of the indicators of a pathological condition in the urinary tract is the presence of more than 100,000 (10^5) bacteria/ml of urine. Most laboratory protocols for the analysis of a urine culture call for some kind of quantitative measure of the bacterial density and identification of the significant potential pathogens.

ORGANISMS FREQUENTLY ENCOUNTERED

The most commonly encountered bacteria from urine cultures include the gram-negative enteric bacteria, the streptococci, and the staphylococci. *Neisseria, Lactobacillus,* and yeasts are also often isolated. The importance of obtaining a good, clean, midstream sample cannot be over emphasized. Failure to carefully cleanse the area or to take a proper midstream sample could result in a sample that is contaminated with normal flora of the skin and associated genital organs (*S. epidermidis, Lactobacillus,* and diphtheroids) and could result in a course of therapy that is not necessary.

1. The lactobacilli are slender, gram-positive rods that produce lactic acid as a product of fermentation. They are catalase negative and are part of the normal flora of the female urogenital tract, particularly the vagina.

2. Yeasts can be identified easily by their typical large size (they are eukaryotic), their characteristic formation of daughter cells by budding, and occasionally by their "yeasty" odor. Since several of the yeasts can grow on Thayer Martin agar and do produce a weak oxidase reaction, it is imperative to do a gram stain on all Thayer-Martin colonies.

3. The gram-negative rods that infect the lower urogenital tract are usually introduced from the intestinal tract. Since we will cover this group in detail in the next exercise, you should save any gram-negative rods that are isolated. They may be initially screened by noting if they ferment lactose and are oxidase positive.

SAMPLE COLLECTION

A clean midstream urine sample is one that is not contaminated with normal flora of the lower urogenital tract or the external genitalia. It must be carefully obtained. The patient is asked to cleanse the external genitalia and to void a portion of urine before collecting the sample. The initial voiding is necessary to rinse bacteria and cellular debris from the urethra.

SEMI-QUANTITATIVE (CALIBRATED LOOP) ESTIMATION OF NUMBERS OF BACTERIA IN URINE

The traditional method of determining the bacterial count is to perform a standard plate count on the urine sample (see Exercise 19). This is accomplished by making pour plates of a series of tenfold serial dilutions of the urine. Since the laboratory usually is only interested in determining if the count exceeds 100,000 bacteria/ml, pour plates of dilutions up to only 1: 1,000 are usually made. Pour plates, however, are not suitable for rapid detection of pathogens; it is also necessary to make isolation streak plates of the sample on differential or selective media. The sample may either be streaked out directly or after mild centrifugation to concentrate the cells in a smaller volume of urine.

A more commonly used alternative procedure to the pour plate is a quantitative technique that allows for the simultaneous estimation of numbers and the direct isolation of pathogens on diagnostic and selective media. This alternate technique is known as the calibrated-loop procedure. A calibrated loop is one that has been carefully calibrated to hold a drop of urine of a specific volume, usually 0.01 ml or 0.001 ml. The loop is dipped into the urine sample and a single streak is made down the center of the plate being used for isolation. This streak is then cross-streaked for isolation of any bacteria that may have been deposited. Any medium may be used, allowing simultaneous isolation of pathogens and estimation of numbers.

Studies have established that when the loops are properly calibrated, the numbers of bacteria in the urine are proportional to the extent of growth down the center streak. If the growth extends more than three fourths of the distance, the urine is considered to have more than 100,000 bacteria/ml and is considered to be evidence of a pathogenic condition. When this occurs, the dominant organism, as well as any suspected pathogens, must be isolated and identified. A more precise estimate can be obtained by counting the colonies that appear on the streak and multiplying that number by 1,000 (if a 0.001 ml calibrated loop was used) to give the number of bacteria per ml. (What would you have to multiply your count by if a 0.01 ml calibrated loop were used?)

An even more rapid method has been developed by Wampole Laboratories. It is the Bacturcult system—a sterile, disposable, plastic tube coated with special nutrient-indicator medium for the rapid detection of bacteriuria and presumptive identification of the causative agent (see Figure 31.1). A urine sample is incubated in the Bacturcult. Following incubation, a counting strip is placed around the tube and the number of colonies within the circular area is

Table 31.1   Bacturcult Results

Average Number of Colonies Within the Circle	Approximate Number of Bacteria per ml	Diagnostic Significance*
less than 25	less than 25,000	negative bacteriuria
25 to 50	25,000 to 100,000	borderline**
more than 50	greater than 100,000	positive bacteriuria

* The final diagnosis of urinary tract infection is dependent upon the clinician's judgement.
** Additional testing recommended.
Source: Wampole Laboratories

determined. The results are interpreted as shown in Table 31.1.

The presumptive identification of the bacteria is dependent upon the color change that occurs in the Bacturcult due to the action of the organisms on the lactose, urea, and phenol red present in the medium. This enables the differentiation of the organisms into three groups as follows:

Group I—*E. coli, Citrobacter, Enterobacter*: yellow.

Group II—*Klebsiella pneumoniae, Staphylococci, Streptococci*: rose to orange.

Group III—*Proteus, Psuedomonas*: purplishred (magenta).

Mixed cultures do not necessarily produce clear results, and additional testing should be performed. The colonies present in the Bacturcult can be used for inoculation of further test media.

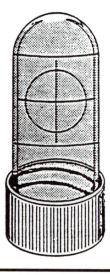

Figure 31.1   Bacturcult

MEDIA USED AND THEIR REACTIONS

1. **Blood agar.** Many of the streptococci isolated will be enterococcus and will appear to be nonhemolytic (gamma-hemolysis). If the sample contains large numbers of other bacteria, the detection of the streptococci may be difficult. Refer to Exercise 30 for a complete description of the reactions on blood, chocolate, and mannitol-salt agars.

2. **Thayer-Martin agar.** The chocolate agars, including Thayer-Martin agar, are used to detect pathogenic neisseriae. Although most bacteria can grow well on chocolate agar, the selectivity of Thayer-Martin agar will ensure isolation of *N. gonorrhoeae*, should it be present.

3. **Mannitol-salt agar.** This agar is occasionally used for the detection of *S. aureus*, although when *S. aureus* is present it can usually be detected on the blood agar streaks.

4. **MacConkey's agar.** MacConkey's agar is a medium that is used for the selection and differentiation of the enteric gram-negative rods associated with the intestinal tract. It contains crystal violet to suppress the growth of gram-positive bacteria. Other components of the medium, including the sugar lactose and a pH indicator (neutral red; red at 6.8, yellow at 8.0), allow differentiation between lactose-fermenting and lactose-nonfermenting gram-negative rods. Lactose-positive bacteria will produce red colonies with a distinctly darker red zone around the colony (due to the color change of the pH indicator), while lactose-nonfermenting bacteria will produce colonies that are light pink or colorless after 24-hour incubation. Whenever the gram-negative rods comprise a significant proportion of the isolated colonies, identification is required, especially if the total count exceeds 100,000/ml.

5. **EMB.** EMB (Eosin Methylene Blue) agar is a selective and differential medium for the detection of the gram-negative enteric bacteria. Organisms which utilize lactose and/or sucrose will appear as blue-black colonies with a greenish metallic sheen, whereas coliform organisms which do not utilize the lactose and/or sucrose form mucoid, pink colonies. The presence of a green metallic sheen on EMB agar is especially diagnostic for the presence of *Escherichia coli*.

OBJECTIVES

This laboratory exercise will introduce you to new procedures that are commonly used in the bacteriological examination of urine, including quantitative techniques for the estimation of the numbers of bacteria in urine. To complete this exercise you will:

1. Understand and explain what a serial dilution is and what it is used for.

2. Explain and differentiate between pour plates and streak plates and discuss their advantages and disadvantages relative to each other.

3. Know what a calibrated loop is and what its applications are.

4. Obtain an understanding of the proper protocol for bacteriological analysis of urine samples.

5. Discuss the importance of a properly obtained urine sample.

6. Achieve some appreciation for the normal and pathogenic flora of the lower urogenital tract.

7. Obtain an understanding of the application of rapid indicator systems for presumptive screening of urine samples.

MATERIALS NEEDED FOR THIS LABORATORY

NOTE: You may not be required to complete all of the procedures listed in the next section of this exercise. For example, your laboratory instructor may have some groups do standard plate counts and others do the calibrated-loop count.

1. Urine samples. Urine samples may be provided for you, or you might be asked to obtain your own (collection bottles and instructions will be provided). It is important that urine samples be processed as soon as possible following collection. If this is not possible, the sample should be refrigerated until ready for use.

2. Gram stain reagents.

Standard Plate Count

3. Three sterile 9.0 ml saline dilution blanks.

4. Four plate-count agar talls, melted and held at approximately 50°C.

5. Four sterile petri plates.

6. Sterile pipettes, 1.0 ml.

Calibrated-Loop Count

7. Calibrated loop.

8. One plate-count agar plate.

Characterization of Known Organisms

9. Known cultures: Representative cultures, including some of the following will be available either as broth cultures for your inoculation onto the designated media, or for observation of their growth on the differential media.
    Staphylococcus aureus
    Staphylococcus epidermidis
    Streptococcus faecalis
    Lactobacillus sp.
    Pseudomonas aeruginosa
    Escherichia coli
    Saccharomyces cerevisiae

10. Media (1 each per known culture assigned by your instructor)
    Blood agar
    MacConkey's agar
    Thayer-Martin agar
    Mannitol-Salt agar
    EMB agar
    NOTE: These may be prepared as biplates or individual plates.

11. Reagents to complete screening tests:
    Hydrogen peroxide
    Coagulase plasma
    Oxidase reagent

Identification of Urinary Tract Organism(s)

12. Media (1 per urine sample)
    Blood agar
    MacConkey's agar
    Thayer-Martin agar
    Mannitol-Salt agar
    EMB agar

13. Reagents to complete screening tests:
    Hydrogen peroxide
    Coagulase plasma
    Oxidase reagent

Use of Bacturcult

14. Bacturcult culture tube (1 per urine sample)

LABORATORY PROCEDURES

Refer to **Figure 31.2** for an overview of the procedures to be performed on the urine sample.

1. Perform gram-stain of urine sample. Record the results on the Laboratory Report Form.

Standard Plate Count (Refer to flow chart in Exercise 19)

1. Obtain

    a. four sterile petri plates and label them "undiluted," "1:10," "1:100," and "1:1000."

    b. three 9.0 ml dilution blanks containing sterile saline and label them "1:10," "1:1 00," and "1 :1 000."

    c. four plate-count agar tails, melted and held at 50°C.

2. With a sterile 1.0 ml pipette, transfer 1.0 ml of urine into the first dilution blank and transfer 1.0 ml of urine into the first tube of melted agar ("undiluted").

3. Dispose of the pipette as indicated by your instructor. Do not use a pipette more than once.

4. Mix the agar by gently rotating the tube. *Do not shake vigorously.* Pour the seeded medium into the first plate and allow it to solidify.

5. Mix the seeded dilution blank by vigorous shaking or swirling. Transfer 1.0 ml from the first blank to the second and 1.0 ml to the second tube of melted agar.

6. As before, mix the medium gently, pour into a petri plate and allow to solidify.

7. Similarly, mix the second dilution blank and make the transfers to the next blank and melted agar tubes.

8. Continue until all four plates have been poured. You should have one plate each for undiluted urine (first plate), 1:10, 1:100, and 1:1000.

9. Incubate at 37°C for 24 hours. If you are unable to perform the count at that time, the

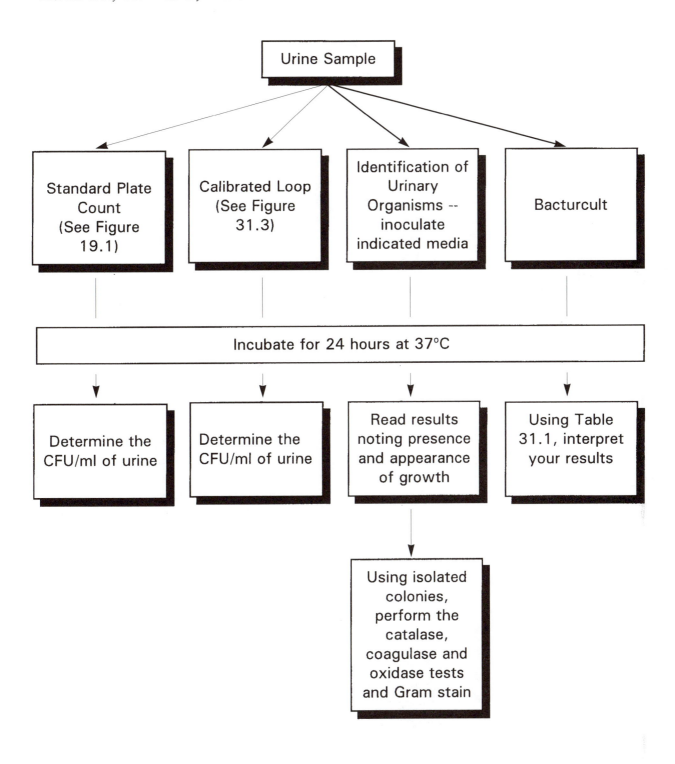

Figure 31.2   Urine Testing Procedure

plates should be refrigerated until your next laboratory period.

10. Determine the CFU (colony forming units) per milliliter of urine. Remember to take into account the dilution of your urine sample.

11. Record the results on the Laboratory Report Form.

Calibrated Loop Count

1. Follow the flow diagram in **Figure 31.3**. Carefully clean the loop, following any special instructions for cleaning and flaming that might be provided by your instructor.

2. Dip the loop into the urine sample so that the loop is just below the surface of the sample. No part of the stem should be in the urine (urine adhering to the stem will cause the counts to come out higher than they should).

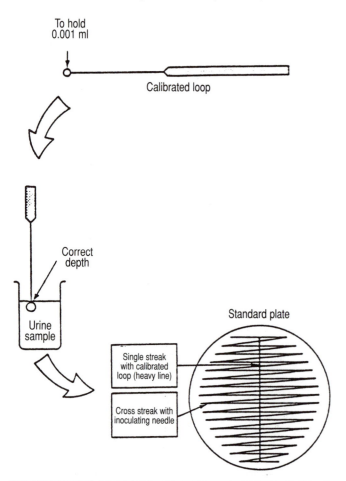

Figure 31.3   Calibrated Loop-Streak Procedure

3. When you remove the loop from the urine, the loop will be filled with a droplet of sample. If the loop has been properly calibrated and correctly handled, it will contain about 0.001 ml.

4. Make a single streak down the center of the plate by touching the filled loop to the top center and quickly sliding the loop down the bottom center of the plate. Cross-streak over the first streak to isolate colonies. Use your regular loop or needle for this isolation streak.

5. Incubate at 37°C for 24 hours. If you are unable to perform the count at that time, the plates should be refrigerated until your next laboratory period.

6. Determine the number of bacteria per milliliter of urine. Remember to take into account the size of the original inoculum. Compare the results obtained with the calibrated loop with those of the standard plate count.

7. Record your results on the Laboratory Report Form.

Characterization of Known Organisms

1. Following the chart in Figure 31.2, inoculate each known organism assigned to you by your instructor on each of the following media. Be sure to streak each for isolation.
   Blood agar
   MacConkey's agar
   Thayer-Martin agar
   Mannitol salt agar
   EMB agar

2. Incubate all plates at 37°C until the next laboratory period.

3. Perform the catalase, coagulase and oxidase tests on isolated colonies. Refer to Exercise 14 as necessary for the required procedures.

4. Complete the table on the Laboratory Report Form for each known culture tested.

Identification of Urinary Tract Organism(s)

1. Following the chart in Figure 31.2, inoculate the urine sample on each of the following media. Be sure to streak each for isolation.
   Blood agar
   MacConkey's agar
   Thayer-Martin agar

Table 31.2   Dichotomous Key—Bacteria Commonly Encountered in Urinary Tract

I.	Gram-positive bacteria	
	A. Catalase positive	
	1. Coagulase positive	*S. aureus*
	2. Coagulase negative	*S. epidermidis*
	B. Catalase negative	
	1. Cocci	*S. faecalis*
	2. Bacilli	*Lactobacillus* sp.
II.	Gram-negative bacteria	
	A. Oxidase positive	*P.aeruginosa*
	B. Oxidase negative	*E. coli*
III.	Typical yeast morphology	*S. cerevisiae*

Note: This key is limited to those tests necessary for the identification of the bacteria used in this exercise. Additional organisms would, of course, require additional tests. Also, a good microbiologist would insist on some confirmatory tests for each isolate.

Mannitol salt agar
EMB agar

2.   Incubate all plates at 37°C until the next laboratory period.

3.   Perform the catalase, coagulase and oxidase tests on isolated colonies. Refer to Exercise 14 as necessary for the required procedures.

4.   Record your results on the Laboratory Report Form. Use Table 30.2 and the results of the known organisms to assist with your identification.

Use of Bacturcult

1.   Label the Bacturcult tube.

2.   Fill the tube almost to the top with the urine sample.

3.   Immediately pour the urine out of the tube, allowing the fluid to drain for several seconds.

Place the lid securely back on the Bacturcult tube.

4.   Loosen the cap on the Bacturcult tube by turning it counterclockwise for one-half turn.

5.   Incubate the tube with the cap down at 37°C for 24 hours. If you cannot read the results at 24 hours, refrigerate the Bacturcult until your next laboratory session.

6.   Place the counting strip around the tube over an area of even colony growth and count the number of colonies within the circle. Using Table 31.1, determine the approximate count per ml and the diagnostic significance.

7.   Observe the color of the medium. Which presumptive group would this indicate?

8.   Record the results on your Laboratory Report Form.

NOTES ——

Name ——————————————

Section ——————————————

LABORATORY REPORT FORM

EXERCISE 31
URINARY TRACT CULTURES

What was the purpose of this exercise?

1. Describe morphology and gram-reaction of organisms seen in the Gram-stain of the original urine sample.

2. Complete the results of the Standard Plate Count and Calibrated-Loop Count of the urine samples.

Sample Source	CFU/ml of Urine	
	Pour-Plate Count	Calibrated-Loop Count

3. Complete the following for the known organisms observed:

ORGANISM	Observed Growth (presence or absence, appearance)				
	Blood Agar	MacConkey Agar	Thayer-Martin Agar	Mannitol-Salt Agar	EMB Agar
Staphylococcus aureus					
Staphylococcus epidermidis					
Streptococcus faecalis					
Lactobacillus sp.					
Pseudomonas aeruginosa					
Escherichia coli					
Saccharomyces cerevisiae					

ORGANISM	Test Reaction			Morphology	Gram Reaction
	Catalase	Coagulase	Oxidase		
Staphylococcus aureus					
Staphylococcus epidermidis					
Streptococcus faecalis					
Lactobacillus sp.					
Pseudomonas aeruginosa					
Escherichia coli					
Saccharomyces cerevisiae					

4. Complete the following for the urine culture tested:

Urine Sample	Observed Growth (presence or absence, appearance)				
	Blood Agar	MacConkey Agar	Thayer-Martin Agar	Mannitol-Salt Agar	EMB Agar

Urine Sample	Test Reaction			Morphology	Gram Reaction
	Catalase	Coagulase	Oxidase		

Identification of organism:

5. Complete the following with the Bactercult results:

Urine Sample	Number of Colonies in Circle	Approximate No. of Bacteria/ml	Diagnostic Significance	Color of Medium	Presumptive Group

How do the results from the Bactercult compare with those achieved with direct count and traditional media/tests?

QUESTIONS:

1. Explain how a clean, midstream or clean-catch urine sample is obtained.

2. What is bacteriuria?

3. List four commonly encountered pathogens associated with lower urinary-genital-tract infections:

4. Why are relatively few microorganisms found in the urine of a healthy individual?

5. Why are females more prone to the development of urinary tract infections than males?

6. What would be the effect of a urine sample being left in a patient's room for several hours before being transported to the laboratory?

GASTROINTESTINAL TRACT CULTURES

Bacterial diseases of the gastrointestinal tract can take many forms—from moderately discomforting "stomach flu" to severe diarrhea that could result in life- threatening loss of fluids and electrolytes. Almost every genus of the gram-negative enteric bacteria have species that can be pathogenic. Two of the more prevalent ones are *Salmonella* and *Shigella*, but, as we have seen, even some strains of *Escherichia coli* can cause severe hemorrhagic colitis. Of course, all of these bacteria, whether usually considered pathogenic or not, are to be considered opportunistic pathogens that will result in severe bacterial infections when introduced to parts of the body that they are not usually associated with.

BACKGROUND

Some nonenteric gram-negative rods, such as certain species of *Pseudomonas* and *Vibrio*, are frequently encountered with the enteric bacteria. One of them, *Pseudomonas aeruginosa*, is usually thought of as nonpathogenic, but can be a very serious pathogen in immunologically compromised hosts (for example, AIDS patients and the elderly) and in burn patients. *P. aeruginosa* is metabolically very versatile and resistant to many of the commonly used antibiotics. It produces a characteristic blue-green pigment in many kinds of growth media. *Vibrio* is a genus of gram-negative, curved rods. Most of the species in the genus are saprophytes, especially in marine environments. Two notable exceptions include *V. cholerae*, which causes cholera, and *V. parahaemolyticus*, a species that is associated with food poisoning involv-

ing seafoods. Both *Pseudomonas* and *Vibrio* are oxidase positive and lactose negative.

The intestinal tract also provides an ideal growth environment for many other kinds of bacteria, including several kinds of enterococci and many anaerobes. Such a large and diverse assembly of bacteria presents some unique problems when it becomes necessary to detect pathogenic species that might be present in very small numbers. Several procedures using highly selective media and isolation or enrichment media have been developed to help resolve this problem.

STRATEGIES FOR STOOL CULTURES

Enrichment cultures. When some members of a bacterial population are present in only a very small proportion of that population, it is necessary to use enrichment media for their isolation. The medium chosen is one that contains selective inhibitors that will inhibit the growth of all bacteria except the species that is to be isolated. The objective of the procedure is to increase the likelihood of isolating the pathogenic species on a streak plate—something that would be very difficult without first enriching the population for the pathogen as normal flora would tend to overgrow the plate.

Selective media. Most of the media used for isolation of the enteric bacteria contain inhibitors to suppress the growth of nonenteric organisms. The two most commonly used inhibitors are the bacteriostatic dyes and bile salts. The dyes inhibit the growth of gram-positive bacteria, and the bile inhibits those bacteria that are not normally found in the intestinal tract (Why?).

Differential media. Many of the differential media used to identify the gram-negative rods do so on the basis of fermentation reactions. When acid is produced, a pH indicator in the medium changes color; a good rule of thumb to follow is that if the colony is a different color than the medium it ferments the sugar. Most of these media use lactose or sucrose as the indicator sugar because virtually none of the pathogenic forms (with one or two notable exceptions) ferment one or the other of these sugars. Therefore, any colony that remains the same color as the medium should be considered suspect and identified. Several media also use additional selective or differential devices. Most of the media commonly used is both selective and differential.

Multitest media. In addition to the plated media, there are several tubed media that have differential properties and can be used to perform more than one test on an isolate. One of the most common is triple sugar iron agar (TSI). It contains three sugars and a pH indicator. The sugars are lactose, sucrose, and glucose, but the lactose and sucrose concentration is ten times (10X) greater than the glucose concentration. When this medium is correctly inoculated, bacteria that ferment glucose only will cause the medium to turn yellow in the bottom of the tube and red at the top (why?). Bacteria that ferment either lactose or sucrose will cause a yellow color to develop throughout the tube. If gas is also produced, bubbles will appear in the agar, causing it to "break" into pieces. This medium also contains iron salts that will turn black when hydrogen sulfide is produced. TSI can, therefore, be used to determine fermentation reaction of three sugars and can test for the production of hydrogen sulfide.

Summary. A typical strategy for the bacterial analysis of a stool culture (or any culture that may contain gram-negative rods) is as follows:

1. Enrichment: Use of a medium that will enhance the growth of the bacterium you want to isolate so that it becomes a prominent member of the population and can be more easily isolated on a streak plate.

2. Selection and differentiation: Use of appropriate media that inhibit the growth of nonenteric bacteria and which differentiate between those

that ferment lactose or sucrose and those that do not.

3. Biochemical identification: Identification of the basis of biochemical reactions, including those determined on certain multitest media.

MEDIA USED AND THEIR REACTIONS

It is especially important that you study the growth patterns and colony morphologies of the reference cultures on the media listed below. Subtle differences in color, colony margin, consistency, and so on can be very important diagnostic leads.

1. **Selenite F broth.** This is a highly selective medium that is used to enhance the growth of the genera *Salmonella* and *Shigella*. The liquid medium is inoculated with a swab saturated with the stool sample and incubated for 12 to 24 hours. Tetrathionate broth is another example of this type of enrichment medium.

2. **Alkaline peptone medium.** Alkaline peptone medium (APM) is used to enrich samples for *Vibrio*. It contains a high salt concentration (3.0%) and a high pH (about 8.0).

3. **MacConkey's agar.** This medium was discussed in Exercise 30. The combination of bile salts and bacteriostatic dyes makes the medium selective for the enteric bacteria. Lactose and pH indicators allow for differentiation between lactose-fermenting and nonfermenting rods.

4. **Hektoen enteric agar.** This medium is highly selective and differential. Organisms that ferment lactose will produce yellow to pink colonies while the nonfermenters will be smaller and green in color. This medium is very reliable and is one of the most commonly used today.

5. **SS agar.** Salmonella-Shigella (SS) agar is selective for these two genera. However, like most such media, the selectivity is not absolute (*Pseudomonas* grows fairly well on it) and gram stains and confirming biochemical tests are always required. *Salmonella* and *Shigella* colonies are light pink, with *Salmonella* occasionally producing a colony with dark centers. The occasional lactose fermenter that does grow on SS agar will produce a distinctly more red colony.

6. **Triple sugar iron agar.** TSI was discussed previously. A yellow color indicates fermentation of one of the sugars. When only glucose is fermented, the deepest part of the medium (the butt) will be yellow, while the part of the medium that is exposed to air (the slant) will be red. The red color appears when glucose is depleted, forcing the bacteria to use the protein components of the medium for growth. When either, or both, of the other sugars is used, the entire tube will be yellow because the concentration of the sugar is large enough so it does not become depleted. As noted above, the iron will react with any hydrogen sulfide produced and cause the formation of a black precipitate in the deep part of the butt. Another medium, Kligler's iron agar, is very similar to TSI except that it has only two sugars (glucose and lactose).

7. **SIM agar.** SIM agar is a medium that can be used to determine sulfide (S) production, indole (I) production, and motility (M). The medium is soft enough so motile bacteria can swim through it, producing a very hazy and cloudy central stab. Nonmotile bacteria produce a sharp and clear stab as they do not move away from the site of inoculation. The sulfide is detected by formation of a black precipitate, and the indole is detected by the addition of Kovac's reagent.

8. **TCBS agar.** TCBS stands for *thiosulfate citrate bile sucrose agar*. It is highly selective for the genus *Vibrio* because of the high pH (about 8.0) and high salt content. It differentiates between sucrose-fermenting and nonfermenting bacteria. The medium can be used for the isolation of Vibrio cholera (yellow colonies) and *V. parahaemolyticus* (green colonies). It is often included in stool-culture protocols during the summer months, when *V. parahaemolyticus* is most likely to be encountered.

RECOMMENDED PROTOCOL FOR STOOL CULTURES

Day 1. Inoculate at least one enrichment medium and at least three selective and differential media by streak plate. Some workers recommend that half of the plates be used for the initial streak, the other half used for streaking the enrichment cultures (See Figure 32.1). The recommended media include:

a. Enrichment media:
Selenite F or tetrathionate broth
Alkaline peptone medium

b. Isolation media:
MacConkey's agar
TCBS agar
Hektoen enteric agar
SS agar

Day 2. Examine the plates previously inoculated and make streak plates of the enrichment culture.

Day 3. Examine all plates for the presence of lactose nonfermenting bacteria. Also examine the TCBS plates for typical Vibrio colonies (yellow or green). Examine all reference cultures. Initiate any biochemical identifications that are appropriate, including TSI, SIM, and/or KIA media.

THE IMViC TEST

The IMViC test is commonly used in the bacterial analysis of water to differentiate rapidly between *Escherichia coli* and *Enterobacter aerogenes*. It is not as commonly used in clinical laboratories because other screening tests have proven more useful. Nevertheless, virtually all screening tests for the gram-negative rods use these reactions (in addition to others) in their protocol. The biochemical tests comprising the IMViC test are Indole (I), Methyl Red (M), Voges-Proskauer (V), and Citrate (C). Note that the i is simply inserted as a phonetic device. See Exercises 12 and 13 for discussions of each of these test procedures. Compare the results for *Escherichia coli* and *Enterobacter aerogenes* for these tests.

LABORATORY OBJECTIVES

Stool cultures are unique because, more so than with any other type of culture, the selection and enrichment for pathogens is an important part of the procedure. In this exercise you will need to:

1. Understand the enrichment process and why it is needed.

2. Understand the principles involved in the use of selective and differential media.

3. Correctly identify lactose and sucrose-fermenting bacteria on the appropriate media.

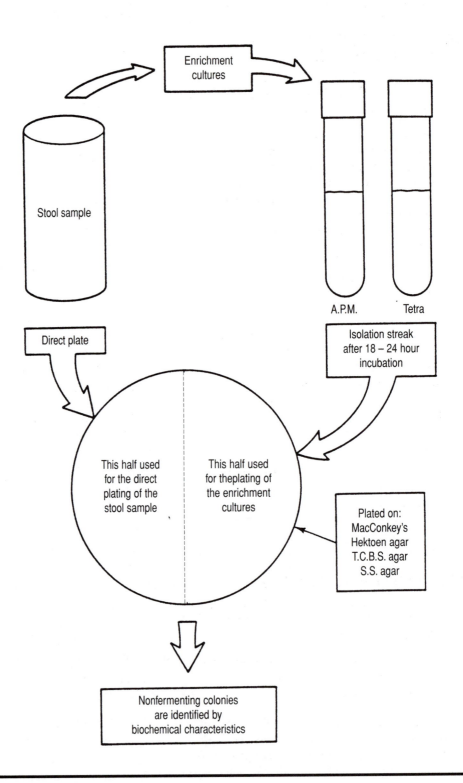

Figure 32.1

4. Understand the procedures and strategies employed in the analysis of stool cultures.

MATERIALS NEEDED FOR THIS LABORATORY

1. Stool specimen. This may be an actual stool sample or a mixed broth sample containing common enteric organisms.

2. Known cultures: Representative cultures, including some of the following, will be available either as broth cultures for your inoculation onto designated media, or for observation of their growth on the differential media.
   Escherichia coli
   Enterobacter aerogenes
   Proteus vulgaris
   Pseudomonas aeruginosa
   Vibrio parahaemolyticus
   Salmonella typhimurium
   Shigella sp.

3. Media needed for each known culture assigned and for each stool sample:
   Selenite F broth
   Alkaline peptone medium
   MacConkey's agar
   TCBS agar
   Hektoen enteric agar
   SS agar

4. For each known culture assigned and for each isolate:
   TSI agar slant
   SIM agar tall

5. Gram stain reagents.

LABORATORY PROCEDURES

1. Inoculate Selenite F broth and Alkaline peptone medium with a swab saturated with the stool sample.

2. Perform primary isolation of the stool sample of the following agars:
   MacConkey's agar
   TCBS agar
   Hektoen enteric agar
   SS agar
   Use only half of each plate for the primary isolation. The other half will be used for streaking the enrichment cultures. (See **Figure 32.1**)

3. Prepare streak plates of assigned known organism on the following agars:
   MacConkey's agar
   TCBS agar
   Hektoen enteric agar
   SS agar
   NOTE: Your laboratory instructor may provide you with demonstration plates of the known organisms.

4. Incubate the enrichment broths and agar plates for 24 hours at 37°C. If you cannot streak them for isolation or read the results at 24 hours, they should be refrigerated until your next laboratory period.

5. Record the results from the primary isolations on the agar plates.

6. Streak out the enrichment cultures on the remaining half of the agar plates.

7. Incubate at 37°C for 24 hours.

8. Observe the growth of the known cultures on the varied media.

9. Record the results for the known cultures on the Laboratory Report Form.

10. Inoculate the known cultures into TSI and SIM agar.

11. Incubate at 37°C for 24 hours.

12. Examine all streak plates, study and compare all colony types, and determine if any lactose nonfermenting bacterial colonies developed on the stool culture streak plates.

13. Inoculate any suspect colonies into TSI and SIM agar.

14. Read and record the results for the known cultures in the TSI and SIM agars.

15. Write a report summarizing your analysis of the stool cultures. It should include:
    a. A description of the known cultures on each of the media used.
    b. A discussion on the use of the enrichment procedure.
    c. A summary of the stool culture analysis: presence or absence of potential pathogens and protocol used for identification or screening.

Table 32.1   Dichotomous Key—Bacteria Commonly Encountered in Gastrointestinal Tract Cultures

I.   Oxidase positive
    A.   Ferments sucrose, curved rod   *V. parahemolyticus*
    B.   Does not ferment any sugar, produces green pigment in some media.   *P.aeruginosa*

II. Oxidase negative
    A.   Ferments lactose
        1.   Produces indole   *E. coli*
        2.   Does not produce indole   *E. aerogenes*
    B.   Does not ferment lactose
        1.   Hydrolyzes urea
            a. Produces indole   *P. vulgaris*
            b. Does not produce indole   *P. mirabilis*
        2.   Does not hydrolyze urea
            a. Motile   *S. typhimurium*
            b. Nonmotile   *Shigella sp.*

Note: This key is limited to those tests necessary for the identification of the bacteria used in this exercise. Additional organisms would, of course, require additional tests. Also a good microbiologist would insist on some confirmatory tests for each isolate.

LABORATORY REPORT FORM

EXERCISE 32
GASTROINTESTINAL TRACT CULTURES

What was the purpose of this exercise?

1. Complete the following chart for the known organisms tested:

	Colony Color			
	MacConkey	**TCBS**	**Hektoen Enteric**	**SS**
Pseudomonas aeruginosa				
Proteus vulgaris				
Proteus mirabilis				
Shigella sp.				
Escherichia coli				
Salmonella typhimurium				
Enterobacter aerogenes				
Vibrio parahaemolyticus				

	TSI and/or SIM				
	Butt	Slant	Sulfide	Indole	Motile
Pseudomonas aeruginosa					
Proteus vulgaris					
Proteus mirabilis					
Shigella sp.					
Escherichia coli					
Salmonella typhimurium					
Enterobacter aerogenes					
Vibrio parahaemolyticus					

2. Complete the following for the stool specimen tested, describing the color of the isolated colonies:

Stool Sample	Primary Isolation			
	MacConkey	TCBS	Hektoen Enteric	SS

Stool Sample	Enrichment Isolation			
	MacConkey	TCBS	Hektoen Enteric	SS

3. Complete the following for each stool isolate tested:

Stool Isolate	TSI			SIM		
	Butt	Slant	Sulfide	Sulfide	Indole	Motility

4. Based on your results above, what organisms appear to have been present in the stool sample tested?

QUESTIONS:

1. Why are lactose and sucrsoe the two most commonly used indicator sugars in the isolation of enteric organisms?

2. What is the purpose of culture enrichment?

3. What is bacterial endotoxin?

4. What is meant by *enteropathic E. coli*?

LACTOBACILLUS ACTIVITY IN SALIVA

Tooth decay has been linked to bacteria found in the oral cavity. Bacteria that are part of the normal flora can form plaque and become embedded in lesions on the teeth. Through a combination of acid production by these bacteria and decomposition of tooth structure, dental caries form and increase in size.

BACKGROUND

The formation of lactic acid by bacteria belonging to the genera *Streptococcus* and *Lactobacillus* has been directly linked to the formation of dental caries. These bacteria, growing in saliva and on the surface of teeth, ferment available carbohydrates to produce the lactic acid. Most of these bacteria are able to ferment sucrose (table sugar) and complete its conversion to lactic acid in the plaque deposits on the teeth or in the saliva of the mouth. This fermentation reaction is completed before the sucrose-containing food is completely chewed and swallowed. (Those who have inflamed gums or teeth with dental caries will feel pain as the acids produced by these bacteria irritate the lesions.) The lowered pH increases the solubility of the enamel, resulting in a dental caries. The formation of plaque and the lodging of food and bacteria in tooth lesions create anaerobic conditions that enhance acid production through fermentation. The resulting process is, therefore, self-sustaining once it begins. When these organisms also produce sticky polysaccharide polymers of dextran, the lactic acid is held against the tooth surface as plaque enabling its breakdown of tooth enamel and dentin. This results in increased tooth decay. It should also be noted that sucrose is also important in the formation of the dextran layer.

MEDIA USED

1. **Rogosa SL Agar.** Rogosa SL agar is a low pH and high acetate medium which suppresses most of the oral flora except the lactobacilli. (Is this medium selective or differential?)

2. **Snyder Test Agar.** Snyder test agar provides a simple correlation between the time of acid production and the number of lactobacilli present in the oral cavity. Brom Cresol Green is used as a pH indicator. It turns yellow at a pH below 4.6. At this point, the organisms present have produced enough acid to decalcify the tooth enamel. When only using Snyder Test Agar as an indicator of the cariogenic potential of oral flora, the following interpretation is commonly used:

Caries Potential	Duration of Incubation		
	24 Hours	48 Hours	72 Hours
Marked	Positive	-	-
Intermed.	Negative	Positive	-
Slight	Negative	Negative	Positive
Insignificant	Negative	Negative	Negative

A positive test is considered as a change in color, as compared to an uninoculated control, with green no longer being dominant. A negative test would then be the absence of any color change, or a tube in which there is only slight deviation with green remaining the dominant color.

EXPERIMENTAL DESIGN

Two procedures have been developed to estimate the amount of acid production of the bacteria in the oral cavity. Both procedures require that saliva be added to a selective medium and acid production measured by color changes in the medium.

One approach is to actually count the number of acid producing bacteria present in saliva. The other approach is to assume that acid production is proportional to the number of bacteria and then to merely determine the rate at which the acid is produced. In other words, the more acid, the more bacteria.

In the Rogosa test, direct pourplate counts of the bacteria in saliva are made using Rogosa SL agar. This medium is highly selective for the genera *Streptococcus* and *Lactobacillus*. The number of colonies that develop is assumed to be representative of the bacterial activity in the oral cavity.

In the Snyder test, saliva is added directly to a tube of Snyder test agar and incubated for up to 72 hours. The medium will turn from green to yellow as bacteria produce lactic acid. Since the production of acid is proportional to the number of bacteria, the sooner the medium turns yellow, the more bacteria present in the saliva sample.

Table 33.1   Color Changes and Number of Bacteria Per mL of Saliva

Time of Color Change (Snyder Test)	Susceptibility to Dental Caries	Approximate No. of Bacteria per ml (Rogosa Test)
< 72 hours	low	< 100/ml
< 48 hours	high	< 10,000/ml
< 24 hours	very high	> 10,000/ml

Both of these tests, in theory, measure the same thing—the amount of lactic acid produced by bacteria in saliva. In the Snyder test, an activity (production of acid) of the bacteria is measured, while in the Rogosa test, the actual number of acid-producing bacteria in saliva is determined. Both procedures assume that there is some direct relationship between the activity or numbers measured under laboratory conditions and the actual production of acid on the surfaces of teeth.

The value of these tests lies in the relationship that exists between the production of acid by bacteria and the formation of dental caries. The accuracy and reliability of the tests depend upon correct sampling, accuracy in reading the tests, and an understanding of the limitations of each procedure. Some important variables include such things as the timing of the sampling period (before or after brushing and flossing), early or late in the day (food accumulation), and the nature of the diet (high in sucrose). Also, variation among individuals, especially in their susceptibility to dental caries, often influences test results. Because of these potential variables, it is recommended that saliva be collected before breakfast and before brushing your teeth. For accuracy, it should be repeated several times.

LABORATORY OBJECTIVES

This exercise uses two approaches to determine susceptibility of individuals to dental caries. To complete this exercise, you must

1. Understand the relationship between fermentation and the production of acid in saliva and on the teeth.

2. Understand the procedures used in this exercise to estimate bacterial activity.

3. Understand the relationship between dietary sugar and the activity of streptococci and lactobacilli in the oral cavity.

MATERIALS NEEDED FOR THIS LABORATORY

For each saliva sample, the following are needed:

1. One tube of Snyder test agar. There should also be one tube available as a control.

2. Three tubes of Rogosa SL agar talls.

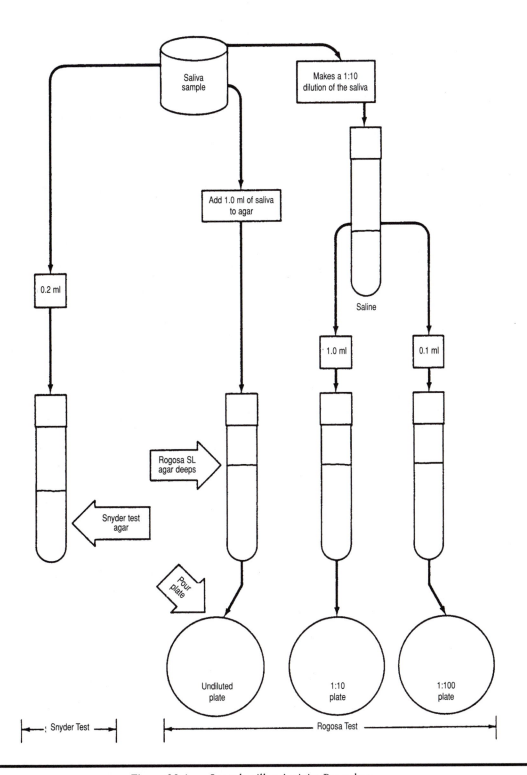

Figure 33.1   Lactobacillus Activity Procedure

3. One tube of 9.0 ml sterile saline.

4. Three sterile petri plates.

5. One sterile tube for the collection of saliva.

6. Sterile pipettes.

7. Small blocks of paraffin for chewing to stimulate saliva production.

LABORATORY PROCEDURES

1. Obtain three sterile petri plates and label them "undiluted," "1:10," and "1:100."

2. Obtain the following media:
   Three tubes of Rogosa SL agar talls
   One 9.0 ml tube of sterile saline
   One tube of Snyder test agar.
   Be sure that the tubes of agar have been melted and cooled to about 50°C.

3. Chew the paraffin block to stimulate production of saliva. Collect at least 2.5 ml of saliva in a sterile test tube.

4. Examine the flow chart shown in **Figure 33.1.**

5. Transfer 1.0 ml of saliva to one of the tubes of Rogosa SL agar. Mix well and pour into the plate labeled "Undiluted."

6. Transfer 1.0 ml of saliva to the tube of sterile saline. Mix well. Pipet 1.0 ml of the dilution to a tube of Rogosa SL agar. Mix well and pour into the plate labeled "1:10." Transfer 0.1 ml of the saline dilution to the remaining tube of Rogosa SL agar. Mix well and pour into the plate labeled "1:100."

7. Transfer 0.2 ml of the **undiluted** saliva into the tube of Snyder test agar. Mix thoroughly and allow to solidify as an agar tall.

8. Incubate all cultures at 37°C.

9. Record the color of the Snyder test agar at 24-hour intervals. Count the colonies that develop on the Rogosa agar at 48 hours.

10. Record all results on the Laboratory Report Form.

LABORATORY REPORT FORM

EXERCISE 33
LACTOBACILLUS ACTIVITY IN SALIVA

What was the purpose of this exercise?

Complete the following information for your saliva sample. Use Table 33.1 to assist in your interpretation of the results.

BACTERIAL COUNT FOR SALIVA AND TIME OF COLOR CHANGE		
	Results	Interpretation
Bacterial count in Rogosa SL agar in 48 hours		
Bacterial count in Rogosa SL agar in 72 hours		
Color of Snyder test agar in 24 hours		
Color of Snyder test agar in 48 hours		
Color of Snyder test agar in 72 hours		

Complete the following chart, using data from other groups in your laboratory section.

Student Number	Time of Color Change (in hours)	Number of Colonies (#/ml)	Number of Fillings or Cavities
1			
2			
3			
4			
5			

QUESTIONS:

1. What correlations, if any, can be observed in the data collected and displayed above?

2. List at least three factors that might affect the results of the Snyder and Rogosa test procedures.

   a)

   b)

   c)

   d)

3. How would you expect the results to change after the use of mouthwash?

4. How would you expect the results to change after eating candy?

5. How would you explain results showing that one of the students in your sample had a very high susceptibility to dental caries on the basis of the Snyder or Rogosa test, yet actually had a very low number of dental caries and/or fillings?

VIRAL POPULATION COUNTS:
PLAQUE COUNTING AND PLAQUE MORPHOLOGY

This exercise involves the plaque formation by viruses on permissive hosts. We will be both counting the number of plaques formed and determining the specific plaque morphology.

BACKGROUND

A *bacteriophage* is a virus that parasitizes bacteria and, like most viruses, lyses the cell that serves as its host. If the cells that the viruses lyse are immobilized in soft agar, a clear area, or *plaque,* is produced. The plaque appears as a circular, clear zone in an otherwise homogeneously turbid or opaque field. The cleared area results, of course, from the lysis of the bacterial cells in (or on) the agar.

Viruses that infect animal or human cells also destroy their host cells, producing plaques when cultured in tissue-culture systems that use tissue monolayer growing on agar surfaces. However, not all animal cells grow well on agar layers; some must be cultured in rotating glass tubes, with the cells growing on the inner surface of the glass, being constantly bathed with medium. When these tissue-culture systems are used, distinct plaques do not develop. Instead, a characteristic change in cell morphology and physiological activity precedes outright cell death. These changes are often referred to as *cytopathic effect* (CPE).

Plaques are analogous to colonies of bacteria growing on a nutrient medium. The bacteria in the agar are the medium on which the viruses grow. Like a bacterial colony, a plaque can be assumed to represent the progeny of a single phage. The number of plaques can be used to count the number of viruses in a suspension. The titration of a phage suspension is simply the determination of t'ne number of plaque-forming units (PFU) present in 1 ml. The most frequently used method for counting phage is to prepare serial dilutions and then to plate the dilutions on lawns of bacterial cells.

Phage plaques show a typical plaque morphology (remember colonial morphology?). When plated on suitable permissive host cells, the morphology of the plaque is characteristic of the phage and the strain of host cells they were grown on. Of course, if the bacterial strain is not a suitable host (nonpermissive) for the virus, plaques will not be observed at all. While the most apparent trait is the diameter of the plaque, other distinctive characteristics can also be observed. These may include:

- Sharpness of the plaque edge
- Presence of a halo
- Relative size of the halo
- Clarity of the plaque

PLAQUE MORPHOLOGY OF T4 AND T4r

Phage T4r is a mutant strain of T4 that produces a dramatically different type of plaque. The *r* stands for *rapid lysis;* phages with this genetic characteristic (the ability to cause rapid cell lysis) produce sharp plaque borders with centers that are clear and free of bacteria. In contrast, T4 will produce a plaque with hazy edges, resembling a halo of reduced turbidity around a partially cleared center. The host strain of bacteria for the T4 family of phages is *E. coli* B.

EXPERIMENTAL DESIGN

The standard plate count, as it is used for counting bacteria (see Exercise 19), cannot be used to count viruses. Reproducible plaque counts and consistent plaque morphology require that the bacteria and viruses be suspended together in the agar during the time the plaques are developing. The most convenient way of doing this is to employ the agar overlay technique for the actual plaque count.

In the *agar overlay technique,* the virus suspension is mixed with the bacteria in about 3 – 5 ml of melted agar. The agar is gently mixed and then poured over an agar layer that has already solidified in petri plates. The lower layer of agar provides a nutrient base for the bacterial and viral growth occurring in the upper layer. Restricting plaque formation to a relatively thin surface layer ensures vigorous virus and bacterial growth and clear, easily seen plaques.

In this exercise you will titrate at least one virus suspension. Your instructor may have you do more that one titration so that you may observe different types of virus plaque morphologies, or varied viral strains may be assigned to alternating students. By varying the strain of phage, differences in the shape of the plaque can easily be observed. Study the flow chart shown in **Figure 34.1** and refer to it as you proceed. If too many phages are plated, crowded or confluent plaques will be produced and it will be difficult, perhaps impossible, to accurately study their morphology.

OBJECTIVES

Viruses are obligate parasites that must be grown on living cells. A virus culture is, in fact, a culture of two organisms: the virus and its host cell. In this exercise you should:

1. Understand the agar overlay technique and the use of serial dilutions for the titration of a virus suspension.

2. Understand what a plaque is and why it is analogous to a virus colony.

3. Learn to distinguish different viruses by their plaque morphology.

4. Understand how viruses cause the destruction of their host cell and appreciate why virus growth in a living organism is usually a serious pathogenic condition.

5. Understand the nature of rapid lysis, as the term is used in this exercise.

6. Realize the biological consequences of plaque formation for the host cells and understand the clinical significance of host-cell destruction *in vivo.*

MATERIALS NEEDED FOR THIS LABORATORY

The following materials are needed for the titration of a single phage.

1. Five prepoured petri plates of TSA. The plates should be warmed in the incubator prior to use.

2. Five tubes of 3 ml soft TSA (0.7% agar), melted and kept at 50°C.

3. Six tubes of 9.0 mi TS broth.

4. 18 – 24 hour culture of *Escherichia coli* B.

5. Suspensions of bacteriophage (T4 or T4r as assigned by instructor). About 2 ml will be needed and should have a titre of about 10,000 phages/ml.

6. Sterile pipettes.

LABORATORY PROCEDURE

1. Be sure to refer to Figure 34.1 as you proceed.

2. Obtain and label 5 petri plates and 5 tubes of soft plating agar. The plates should be labeled with the dilution (1 :10 to 1: 1 00,000) and the phage type and host-cell strain. You will also need 6 dilution blanks containing 9.0 ml of TS broth.

3. Maintain the soft agar in a 50°C water bath to prevent it from solidifying.

4. Transfer about 5 ml of TS broth from one of the tubes to the slant of E. coli B. Rock the slant back and forth to suspend the bacterial cells. Label the remaining five blanks according to the serial dilution series (1 :10 to 1 :100,000).

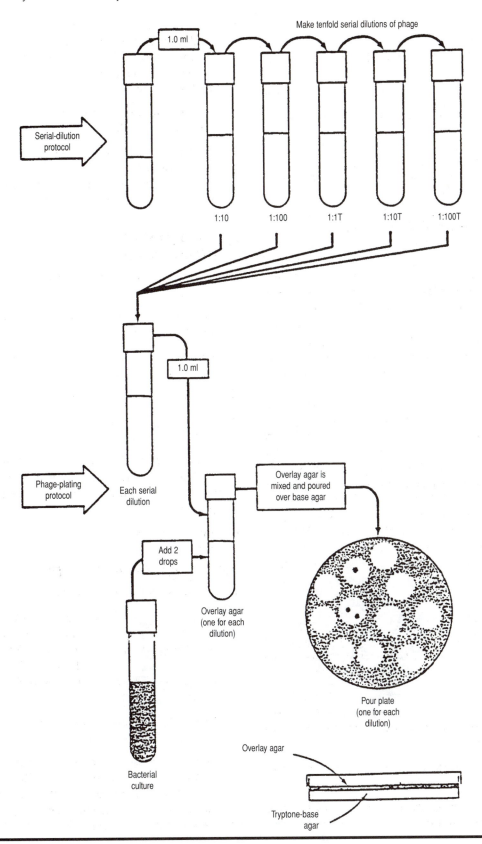

Figure 34.1   Plaque Count Procedure

5. Prepare a serial dilution of the phage suspension. Transfer 1 ml from the undiluted stock suspension to the first dilution blank. Mix well, then transfer 1 ml to the next blank. Repeat these transfers until all dilutions have been made.

6. Transfer two drops of the bacterial suspension into each of the soft agar tubes.

7. Transfer 1 ml from the first phage dilution (1:10) into the first soft agar tube. Mix gently and immediately pour into the appropriately labeled plate.

8. Repeat this procedure until all the dilutions have been plated.

9. Incubate all plates at 37°C for 24 hours. If you cannot complete the exercise after 24 hours incubation, store the plates in a refrigerator until the next laboratory period.

10. Count the plaques on all dilutions with 25 to 250 plaques present. Observe the morphology of the plaques observed (See **Figure 34.2**).

11. Complete the Laboratory Report Form for this exercise.

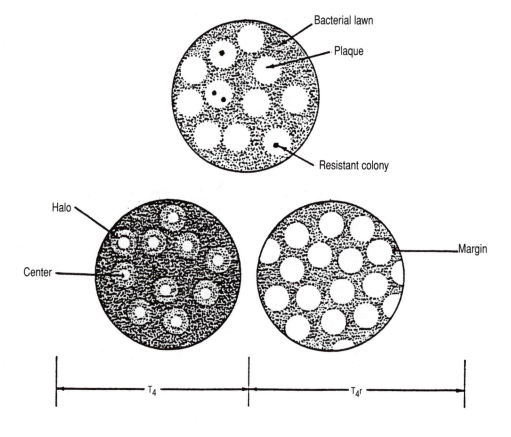

Figure 34.2   Plaque appearance.

Name _____

Section _____

LABORATORY REPORT FORM

EXERCISE 34
VIRAL POPULATION COUNTS

What was the purpose of this exercise?

1. Complete the following table. You should obtain any necessary data from other students in the class and use their plates to determine plaque size and morphology.

RESULTS OF PLAQUE ASSAY

Phage (Host)	Count (PFU/ml)	Size of Center in mm	Size of Halo in mm	Morphology and Margin
T4 (E. coli B)				
T4r (E. coli B)				

NOTE: The plaque counts must be calculated from the number of plaques that formed on the plates. You should determine the number of plaque-forming units in each milliliter of stock suspension (PFU/ml).

2. Did you observe any variation in the plaque morphology of any of these phages? If so, what is the significance of such variation?

3. How does your plaque count compare with the expected count, based on the reported titres of the phage suspension? Explain any differences.

4. Did any of the plaques have colonies growing within the cleared area? If so, what is the significance of those colonies?

QUESTIONS:

1. Describe the agar-overlay technique.

2. What is rapid lysis?

3. Compare the number of plaques on each of the dilutions. Do they agree with the dilution factor? (Does the 1:10 plate have ten times as many plaques as the 1:100 plate?)

FOOD MICROBIOLOGY

Most foods provide an excellent growth medium for bacteria and other microorganisms. The supply of organic matter is plentiful, the water content is usually sufficient, and the pH is neutral or only slightly acidic. The result is food spoilage which implies an economic loss to the manufacturer and a waste of money to the consumer, as well as a threat to health. Many types of food spoilage are aesthetically displeasing, and we refrain from eating the food even though the spoilage does not pose a life-threatening situation. In this exercise we will examine the presence of bacteria on various food products.

BACKGROUND

Microorganisms may be introduced into food at any time along its route from farm to our table. Plants will contain organisms found in soil, water, and air naturally occurring on their surfaces. In addition, organisms may penetrate the surface and be present in deeper tissue. Animals also are going to contain significant number of organisms as part of their normal flora. During the slaughtering process, these organisms may be introduced to the muscle tissue— an area normally free of microorganisms. These food products are then processed in some manner and stored for future use. Here it is possible for equipment and handlers to once again introduce organisms into the food product. Many, if not most, of these organisms are destroyed in the final processing of these products (canning, freezing, pasteurizing, drying, etc.). Even so, some organisms may remain and more will likely be introduced as this product is readied for eating where it again encounters utensils and handlers.

It is these organisms which remain in our food products that can contribute to food spoilage and food-borne illness. To minimize potential foodborne problems, standard plate counts are often performed on food samples by food-processing companies and public health agencies. These determine the overall quality of the food, the cause of spoilage, the existence of proper storage methods, and the potential presence of pathogens.

The standard plate count involves diluting small samples of food, plating it using a general media, incubating, and determining the number of organisms present. Whereas these results are indicative of overall food quality, the information gained is limited. If varied incubation temperatures are used, it is also possible to determine the relative presence of thermophiles or psychrophiles. (Why would this be important?) In addition, various selective and differential media can be used for the detection of specific groups of organisms, such as coliforms (MacConkey Agar), *Salmonella* or *Shigella* (Hektoen enteric or SS agar), and *Staphylococcus aureus* (Mannitol-salt agar). Not only is this information important in determining the overall quality of food products, but it also aids in the development of adequate preservation methods for varied products and in assessing their effectiveness. Bacterial counts of foods are not, however, directly indicative of food safety. It is possible for a product to have relatively high total counts and yet be completely safe to eat. For this reason, total counts are of little value when determining the quality of fermented food products (e.g., yogurt or pickles). Plate counts also do not indicate the presence of microbial toxins in foods. For these reasons, it is important that we remember that the total plate count of a food product is only an indication of the organisms that are present and that can grow under the incubation conditions used.

LABORATORY OBJECTIVES

In this laboratory exercise, you will be performing plate counts to determine the quantity of bacteria present in/on varied food products. At its completion you should be able to

- Perform and evaluate a quantitative analysis of the bacteria in food samples.

- Understand the significance of bacterial plate counts when applied to food.

MATERIALS NEEDED FOR THIS LABORATORY

The following will be needed for each food sample being tested.

1. Plate count agar talls (5). These will need to be melted and then held at 50°C to keep them liquid.

2. Sterile 9.0 ml water blank OR sterile 90.0 ml water blank (1)

3. Sterile 99.0 ml water blank (2)

4. Pipettes, 1.0 ml and 10.0 ml

5. Sterile petri plates (5)

6. Balance

7. Sterile blender or mortar and pestle

PROCEDURE

Refer to the flow chart in Figure 35.1 as you perform this exercise.

1. Obtain five sterile petri plates and label them 1:10, 1:100, 1:1000, 1:10,000, and 1:1 00,000.

2. Obtain one sterile 9.0 ml water blank and two sterile 99.0 ml water blank.

3. Aseptically weigh 1 gram or pipette 1 ml of the food to be tested.

    a. If the food is solid, it should be placed into the blender or mortar. Gradually add the 9.0 ml sterile water blank, mixing the food and fluid thoroughly with the blender or the grinding of the pestle.

    b. If your food sample is liquid, directly add it to the 9.0 ml sterile water blank and mix. This will give you a 1:10 dilution.

NOTE: If 1 gm of the food sample will not provide a representative sample of the product, you may weigh out 10 gm of the food and add it to 90 ml of water. This will also provide a 1:10 dilution.

4. Pipette 1.0 ml of your dilution into the petri plate labeled 1:10 and 0.1 ml of your dilution into the petri plate labeled 1:100. Be careful to avoid airborne contamination.

5. Pipette 1.0 ml of the dilution into the 99.0 ml sterile water blank. Mix thoroughly. Pipette 1.0 ml of the dilution to the petri plate labeled 1:1000 and 0.1 ml to the plate labeled 1:10,000.

6. Pipette 1.0 ml of the 1:1000 dilution into the second 99.0 ml sterile water blank. Mix thoroughly. Pipette 1.0 ml of the dilution to the petri plate labeled 1:100,000.

7. Obtain 4 tubes of melted plate count agar which has been maintained at 50°C.

8. Aseptically pour a melted plate count agar tall into each plate, rotating each carefully to mix the agar and dilution. Allow the media to harden.

9. Incubate at 37°C for 24 to 48 hours.

10. Count the colonies, both surface and subsurface, in your plates. Record the results on the Laboratory Report Form. Use the term "TNTC" to refer to those plates with well over 300 colonies. Select the dilution with a final count of between 30 and 300 colonies and calculate the total plate count per gram(ml) of food.

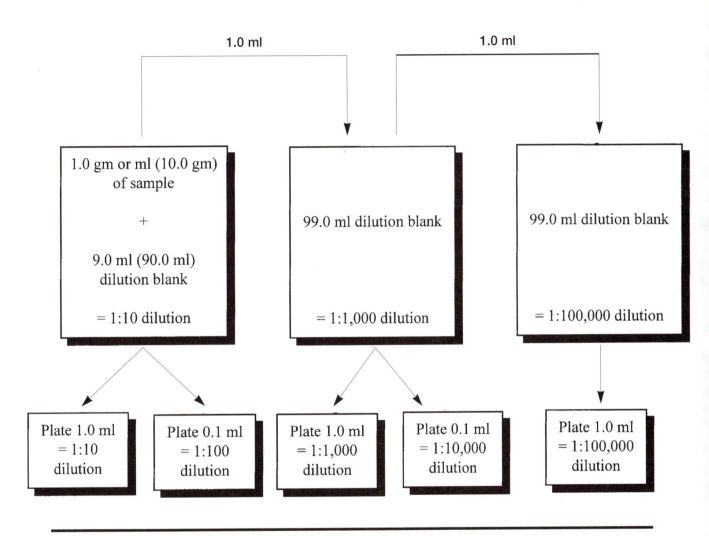

Figure 35.1   Food Testing Procedure

NOTES

LABORATORY REPORT FORM

EXERCISE 35
FOOD MICROBIOLOGY

What was the purpose of this exercise?

What food was tested?

Plate Count results:

Dilution	1:10	1:100	1:1000	1:10,000	1:100,000
Plate Count					

Calculate the CFU/gram or ml of your food sample.

Dilution used   _____

Count results   _____

CFU/gram or ml   _____

Record the results for other food products tested.

FOOD SAMPLE	CFU/gm or ml

QUESTIONS:

1. Does a high standard plate count in food indicate that it must not be eaten? Why or why not?

2. What errors may account for inaccurate plate count results?

3. You have gotten very similar total counts (approximately 10^7 CFU/gm) for both ground beef and potato salad from the deli. How would you interpret these results as to the safety of food consumption?

4. You have a can of soup with bloated ends. You know that you shouldn't eat it as the food is very likely spoiled and decide that you want to see what microorganism(s) is(are) present. What incubation conditions would you use when testing this product?

5. When testing for bacteria in shellfish, MacConkey agar is often used instead of plate count agar. Why would a selective/differential media be chosen? What type of organism is being tested for and why?

WATER ANALYSIS:
STANDARD METHODS AND MEMBRANE FILTER TECHNIQUES

Freshwater supplies are often subjected to the introduction of raw sewage, inorganic and organic industrial wastes, and fecal contamination by all animals, including humans. This may result in the occurrence of dead fish on the beach, or an outbreak of typhoid fever or cryptosporidium in the human population. In this exercise we will be examining the presence of bacteria in standing bodies of water through the utilization of a standard method analysis and membrane filter technique.

BACKGROUND

A multitude of pathogenic organisms can be transmitted through the ingestion of contaminated water. These include *Salmonella* and *Shigella* sp., gastroenteritis and hepatitis viruses, and the protozoa *Giardia, Entamoeba,* and *Cryptosporidium.* To test water for each potential pathogen is not practical. Instead our water safety standards have been based on the presence (or better, the absence) of readily determined, usually nonpathogenic fecal organisms. If a water supply is found to contain any microorganisms of fecal origin, it might also contain pathogens and, therefore, should not be considered safe to consume. Those bacterial organisms that are commonly used as indicators of fecal contamination are referred to as the conform bacteria.

COLIFORMS

Coliforms are those bacteria that have characteristics similar to *Escherichia coli* They are aerobic or facultative, gram-negative, non-spore-forming bacilli which ferment lactose with the production of acid and gas. In addition, they will grow on EMB (Eosin-Methylene Blue) agar where *E. coli* usually produces a green "metallic" sheen. These organisms are usually non-pathogenic, however they are usually found in large numbers in animal feces (as many as 10^6 to 10^9 cells/gm of feces) and survive longer in cold surface water than do most intestinal pathogens. In addition, direct correlations have been shown between the presence of these indicator organisms and the degree of fecal contamination. Another extremely significant reason to utilize these organisms is their relative ease of detection.

The coliform group of bacteria includes (in addition to the *Escherichia*) *Enterobacter, Citrobacter* and *Klebsiella* organisms. These other organisms, however, may also enter the water via soil and vegetation runoff and so are not strictly enteric in origin. Because of this, some prefer to screen for the presence of fecal coliforms, or those organisms which are gram-negative, non-spore-forming bacilli that ferment lactose producing acid and gas at 44.5°C in 24 hours. By using an elevated temperature of incubation, many of the non-enteric coliforms are eliminated.

Recent studies have indicated that the fecal enterococci might serve as more accurate indicator organisms. These gram-positive enteric organisms are less likely to be introduced by nonfecal methods but are not as readily tested for.

BACTERIOLOGICAL EXAMINATION OF WATER

Two methods have routinely been used for the determination of water contamination. The first is the plate count method. This involves plating dilutions of water on nutrient agar and incubating at 20°C and 35°C. It is well understood that this will not indicate the total number of organisms in the water, but only those that can grow under the incubation conditions used. It can, however, indicate the relative number of organisms present in water. This method does not distinguish between environmental and fecal microorganisms so it is not commonly used for determination of bacterial quality of water supplies.

The second major method involves the utilization of indicator organisms, such as the coliforms, which are present in large numbers in fecal material. It should be mentioned at this time that indicator organisms are not always effective in indicating the potential presence of pathogens. More sophisticated testing methods give reason to suspect that enteric pathogens may be present in the absence of indicator organisms due to their increased resistance to the aquatic environment as well as to chlorination of water. In spite of these limitations, indicator organism screening continues to be the most common method of bacteriologic analysis of water. Commonly performed testing protocol include the Standard Methods Analysis and the Membrane Filter Technique. Both of the methods mentioned, however, are somewhat limited in their ability to determine the level of coliform contamination.

MOST PROBABLE NUMBER (MPN) ANALYSIS

The most probable number analysis, or MPN, while not commonly used in clinical applications, is extensively used in public health and environmental microbiology. There are two important advantages of the MPN procedure. First, it is a good screening procedure, useful when a high level of precision is not required. (Do you need to know *exactly* how many bacteria are present, or is it sufficient to know *about* how many are present?) The second advantage is that selective and differential media can be used. This makes it possible to estimate numbers of specific kinds of bacteria, such as coliforms in water. It must be noted, however, that the data obtained from an MPN test is only an estimate of the most *probable* number of bacteria in a given volume of sample.

Table 36.1   Most Probable Number Sets

Set	10 ML of	Volume of Sample Used
First	Double-strength broth	10.0 ml
Second	Single-strength broth	1.0 ml
Third	Single-strength broth	0.1 ml

The MPN test is accomplished by inoculating three sets of tubes with known volumes of sample. Typically there are five or three tubes of media in each of three sets. The first set usually contains 10.0 ml of double-strength broth. Each tube in this set is inoculated with 10.0 ml of sample. Double-strength broth is used so the medium will not be too dilute when you add an equal volume of sample. The second and third sets of tubes contain 10.0 ml of single-strength broth and are inoculated with 1.0 ml and 0.1 ml of sample, respectively. (See Table 36.1)

After a suitable incubation period, the pattern of growth obtained is noted and compared to a table of statistically determined most probable numbers (See Table 36.2). For example if there are five tubes in each set and 3 of the 10.0 ml tubes and 1 of the 1.0 ml tubes showed growth (as: + + +--/+----/-----), the most probable number would be 11 bacteria per 100 ml of sample. The MPN procedure estimates the most probable number of bacteria likely to be present in 100 ml of the sample. The MPN tables give that value for each of the possible combination of growth in the three sets of tubes, from all positive (+ + + + +/+ + + + +/+ + + + +) to all negative (-----/-----/-----). In addition to the probable number present, the table gives the 95% confidence limits for the given test result. This tells us that, for the test example given above, there is 95% assuredness that the actual value is between 4.0 and 29 organisms/100 ml of sample.

STANDARD METHODS ANALYSIS

The Standard Methods Analysis of water is a three step process—the presumptive test, the confirmed test, and the completed test.

1. **Presumptive test.** The presumptive test is an enrichment procedure for the presence of coliforms using a selective medium. Lauryl sulfate tryptose lactose (LST) broth is inoculated with aliquots of the water and incubated. This medium contains lauryl

Number of Positive Tubes			MPN/ 100 ml	95% Confidence		Number of positive tubes			MPN/ 100 ml	95% Confidence	
10.0 ml	1.0 ml	0.1 ml		Lower	Upper	10.0 ml	1.0 ml	0.1 ml		Lower	Upper
0	0	0	< 2	--	--	4	3	0	27	12	67
0	0	1	3	1.0	10	4	3	1	33	15	77
0	1	0	3	1.0	10	4	4	0	34	16	80
0	2	0	4	1.0	13	5	0	0	23	9.0	86
1	0	0	2	1.0	11	5	0	1	30	10	110
1	0	1	4	1.0	15	5	0	2	40	20	140
1	1	0	4	1.0	15	5	1	0	30	10	120
1	1	1	6	2.0	18	5	1	1	50	10	150
1	2	0	6	2.0	18	5	1	2	60	30	180
2	0	0	4	1.0	17	5	2	0	50	20	170
2	0	1	7	2.0	20	5	2	1	70	30	210
2	1	0	7	2.0	21	5	2	2	90	40	250
2	1	1	9	3.0	24	5	3	0	80	30	250
2	2	0	9	3.0	25	5	3	1	110	40	300
2	3	0	12	5.0	29	5	3	2	140	60	360
3	0	0	8	3.0	24	5	3	3	170	80	410
3	0	1	11	4.0	29	5	4	0	130	50	390
3	1	0	11	4.0	29	5	4	1	170	70	480
3	1	1	14	6.0	35	5	4	2	220	100	580
3	2	0	14	6.0	35	5	4	3	280	120	690
3	2	3	17	7.0	40	5	4	4	350	160	820
4	0	0	13	5.0	38	5	5	0	240	100	940
4	0	1	17	7.0	45	5	5	1	300	100	1300
4	1	0	17	7.0	46	5	5	2	500	200	2000
4	1	1	21	9.0	55	5	5	3	900	300	2900
4	1	2	26	12	63	5	5	4	1600	600	5300
4	2	0	22	9.0	56	5	5	5	> 1600	--	--
4	2	1	26	12	65						

Table 36.2   MPN Index and 95% Confidence Limits for Various Combinations of Positive Results When Five Tubes are Used per Dilution.

From *Standard Methods for the Examination of Water and Wastewater,* 18th edition, 1992. American Public Health Association, the American Water Works Association, and the Water Environment Federation. Reprinted with permission.

sulfate, a surface active detergent which inhibits the growth of Gram-positive organisms, and lactose. The presence of acid and gas in 24 to 48 hours is considered presumptive evidence for the presence of coliform organisms. When the presumptive test in performed in this manner, it is simply a qualitative test answer the question, "Are there any coliform organisms present?" Whereas this is important to know, it is even more significant to know the degree of coliform contamination. The presumptive test can be used as a MPN determination by using multiple tubes of LST.

2. **Confirmed Test.** The presence of a positive presumptive test is confirmed with the confirmed test. Brilliant green lactose bile (BGLB) broth is inoculated from a positive presumptive tube and incubated at 35°C. This medium further inhibits the presence of any Gram-positive organisms by the introduction of bile salts and brilliant green. The production of gas within 24 – 48 hours "confirms" the presence of coliform organisms. A variation of the confirmed test that is often performed involves the additional inoculation of EC medium which is then incubated at 44.5°C. The EC broth is also a lactose and bile salts medium which further inhibits non-fecal organisms with its elevated temperature of inoculation. It is, therefore, highly selective for the fecal coliforms.

3. **Completed Test.** The completed test may be performed to further verify the presence of *Escherichia coli* and distinguish its presence from that of *Enterobacter aerogenes* (commonly found in soil and plant runoff). It may involve several testing procedures, including:

a. EMB agar for presence of characteristic growth,

b. NA slant for Gram staining,

c. LST or BGLB broth for confirmation of lactose fermentation,

d. IMViC testing for presence of typical test patterns (see Exercise 32 for further discussion of the IMViC tests).

When the presumptive and confirmed tests show distinctive positive results the completed test is not considered essential for the confirmation of the presence of coliform organisms and so is not performed.

MEMBRANE FILTER TECHNIQUE

The membrane filter technique is widely used for the detection of coliforms in water samples. The testing procedure, which was demonstrated in Exercise 19, is recommended by the American Public Health Association for the testing of water because of the distinct advantages it has over Standard Methods testing. These advantages include:

1. directly counts of either coliform colonies or colonies of other organisms can be made. For example, a direct count of *Salmonella* present in water could be more readily determined;

2. the results can be obtained in a shorter time;

3. larger volumes of water can be tested; and

4. the results are more accurate and reproduced more readily **when** appropriate media and incubation conditions are used. In addition, this testing protocol is less expensive to perform.

Volumes of water are filtered through a membrane filter with a pore size of 0.45 μm. The filter can then be directly placed on either an agar medium or a blotter pad saturated with broth. For example, the filter can be placed on a pad saturated with mEndo MF broth and incubated at 35°C for detection of total coliform counts. If the filter is placed on m-FC broth and incubated at 44.5°C, fecal coliform determinations can be made; incubation on KF streptococcus agar at 35°C will give fecal streptococci values. The use of membrane filters, however, is not without its limitations. In order to determine accurate results, appropriate dilutions of the water sample must be made to give final results of 20 – 200 colonies per plate. In addition, if the water is highly turbid, contains silt or large numbers of bacteria or algae it may clog the filter making accurate determinations very difficult.

LABORATORY OBJECTIVES

In this exercise you will be performing the Standard Methods Analysis on a water sample and observing the use of the Membrane Filter technique. Through this exercise, you should

• Understand the Standard Method analysis of water and the three steps involved in the complete analysis.

- Learn how to use the MPN tables for water quality determination.

- Understand the Membrane Filter technique of water analysis.

- Know the basis for selective and differential media used in water analysis.

- Be able to interpret the results obtained as to overall water quality.

MATERIALS NEEDED FOR THIS LABORATORY

If desired, your instructor may have you bring a water sample from home, a nearby stream, a lake or some other location for testing. You should use a sterile collection bottle (available from your instructor) for water collection. Surface water (lake, pond, stream, etc.) samples should be taken under the surface so as to prevent the introduction of material floating on the surface. If the sample cannot be collected within 1 – 2 hours of testing, keep it refrigerated until your laboratory period.

A. STANDARD METHODS ANALYSIS

1. Water sample

2. Five tubes of Lauryl sulfate tryptose lactose (LST) broth (double strength), 10 ml/tube, containing Durham tube.

3. Ten tubes of Lauryl sulfate tryptose lactose (LST) broth (single strength), 10 ml/tube, containing Durham tube.

4. One tube of Brilliant green lactose bile (BGLB) broth containing Durham tube.

5. One plate Eosin methylene blue (EMB) agar

6. Pipettes, 10.0 ml and 1.0 ml

7. Pipetter

B. MEMBRANE FILTER TECHNIQUE

This procedure may be done as a demonstration.

1. Water sample

2. Media which may include:
    m-Endo Broth MF
    m-FC Broth
    KF Streptococcus Broth

3. Bacteriological membrane-filter assembly, sterile

4. Sterile membrane filters, 0.45 μm pore size

5. 47-mm Petri plate, sterile

6. Absorbent pads, sterile

7. Sterile distilled water, 50 ml

8. Pipette, 5.0 ml

9. Forceps

10. 95% ethyl alcohol

LABORATORY PROCEDURE

A. STANDARD METHODS ANALYSIS

Before starting this procedure, review the protocol as shown in **Figure 36.1**.

1. Obtain and label 5 tubes of double-strength LST broth. Inoculate each with 10.0 ml of your water sample. Be sure to carefully mix the water sample prior to sampling.

2. Obtain and label 10 tubes of single-strength LST broth. Inoculate five tubes with 1.0 ml of your water sample in each. Inoculate the remaining five tubes with 0.1 ml of water sample in each.

3. Incubate all tubes at 37°C for 24 – 48 hours.

4. Following incubation, determine the presence of acid and gas in the tubes. Record you results on the Laboratory Report Form, Part A1. Using Table 36.2, determine the MPN value and confidence levels for your sample.

5. From one of your most dilute positive LST broth tubes, inoculate a tube of BGLB broth. If your water sample was negative at all dilutions, use a positive presumptive test tube from another student to complete the testing procedure.

6. Incubate the tube at 37°C for 24 hours. If you are unable to examine the tube and complete the testing at that time, refrigerate the tube until your next laboratory period.

7. Following incubation, read the results of the BGLB broth. Record the results on the Laboratory Report Form, Part A2.

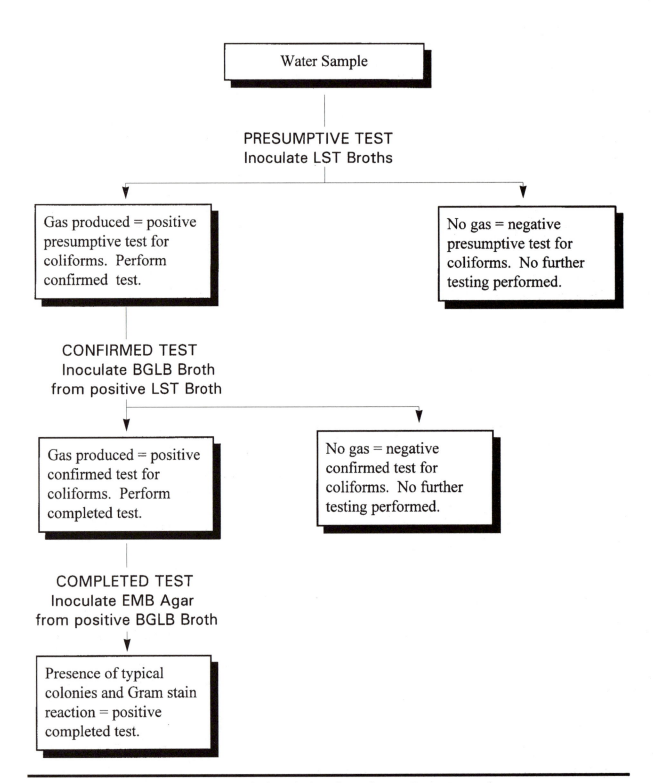

Figure 36.1 Standard Method Analysis of Water Testing Protocol.

8. From a positive BGLB test, streak a plate of EMB agar.

9. Incubate the plate at 37°C until your next laboratory period.

10. Following incubation, examine the EMB plate for the presence of colonies with a green "metallic" sheen. These colonies are positive for the fermentation of lactose. Record the results on the Laboratory Report Form, Part A3.

11. Perform a Gram-stain on a representative colony.

B. MEMBRANE FILTER TECHNIQUE

Preparation of Petri Dish

1. Sterilize a pair of forceps by dipping them in alcohol and then flaming to burn away the alcohol. Aseptically place a sterile absorbent pad in the bottom of each 47 mm Petri plate.

2. Pipet 2 ml of m-Endo MF broth onto the absorbent pad. Replace the lid on the plate.

3. Repeat steps 2 and 3 for other media as instructed.

4. Assemble the membrane filter apparatus as shown in **Figure 36.2**.

5. Aseptically, using sterile forceps, place a sterile membrane on the filter base.

6. Pour 50 ml of sterile distilled water in the receptacle. Pipet 1.0 ml of water sample into the receptacle.

7. Gently turn on the vacuum until the water has been completely drawn through the filter.

8. Using sterile forceps, transfer the filter to the surface of the absorbant pad. Carefully place it on the pad so as to not trap any air under the filter.

9. Incubate the Petri plate at 37°C for 24 hours.

10. Repeat the filtering process, as directed by your instructor, for plating on additional media.

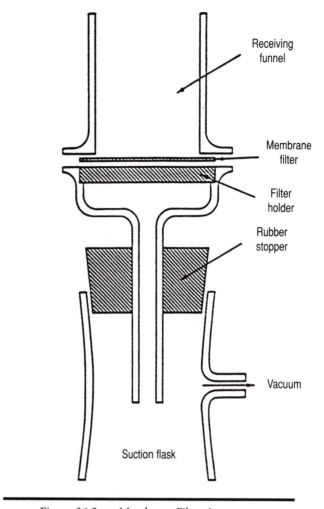

Figure 36.2   Membrane Filter Apparatus

Incubate the plates at 37°C with the exception of the m-FC broth plates which must be incubated at 44.5°C.

11. Determine the number of colonies per plate for each media used. Calculate the number of organisms per 100 ml of your water sample. Record your results on the Laboratory Report Form, Part B.

NOTES

LABORATORY REPORT FORM

EXERCISE 36
WATER ANALYSIS

What was the purpose of this exercise?

A. STANDARD METHODS ANALYSIS

Source of water tested:

1. Presumptive Test.
   Record the number of tubes at each dilution that were positive for the presence of both acid and gas following incubation.
   10.0 m.   _____
   1.0 m.   _____
   0.1m.   _____

   Using Table 36.2, determine the MPN for your water sample.
   MPN   _____
   95% Confidence Limits   _____

2. Confirmed Test.

   Describe the results obtained in the BGLB broth.

   Was the confirmed test positive or negative?

3. Completed test.

   Describe the appearance of colonies on the EMB agar.

   Gram stain results:

         Morphology   _____

         Gram reaction   _____

B. MEMBRANE FILTER TECHNIQUE
   Source of water tested:

   Total coliform count: _____   CFU/100 ml water

   Fecal coliform count: _____   CFU/100 ml water

   Fecal streptococci count: _____   CFU/100 ml water

QUESTIONS

1. Why are laboratory tests for the bacterial pollution of water based on the identification of coliform bacteria and especially *Escherichia coli*?

2. What could be the cause of a positive presumptive test that does not test positive with the confirmed test?

3. Could a sugar other than lactose be used in the media for coliform detection? Why or why not?

4. Can water with a negative presumptive test following 24 hours incubation be considered completely safe to drink?

5. What limitations are there to using the membrane filter technique?

6. Explain why the filtration of contaminated water through a membrane filter does not necessarily guarantee that it is safe to drink.

ALGAL CULTURES

Gleocapsa

BACTERIAL CULTURES

Acinetobacter calcoaceticus
Bacillus cereus
Bacillus megaterium
Bacillus stearothermophilus
Bacillus subtilis
Clostridium sporogenes
Corynebacterium diphtheriae
Corynebacterium xerosis
Enterobacter aerogenes
Enterococcus faecalis
Escherichia coli
Halobacterium salinarium
Klebsiella pneumoniae
Lactobacillus sp.
Lactobacillus delbruckii subspecies *bulgaricus*
Lactococcus lactis
Micrococcus luteus
Moraxella (Branhamella) catarrhalis
Mycobacterium smegmatis
Neisseria gonorrhoeae
Neisseria meningitidis
Proteus mirabilis
Proteus vulgaris
Pseudomonas aeruginosa
Pseudomonas fluorescens
Salmonella typhimurium
Serratia marcescens
Shigella flexneri

Staphylococcus aureus
Staphylococcus epidermidis
Streptococcus sp.
Streptococcus pneumoniae
Streptococcus pyogenes
Vibrio parahaemolyticus

MOLD CULTURES

Aspergillus
Penicillium
Rhizopus

PHAGE CULTURES

Coliphage T4
Coliphage T4r
Host Bacteria: *E. coli* B

PROTOZOA CULTURES

Amoeba
Euglena
Paramecium

YEAST CULTURES

Saccharomyces cerevisiae

SOURCES

Smaller laboratories may find that commercially prepared cultures are often more convenient than keeping stock cultures. This is particularly true of the

reference cultures available in dried disk form from Difco and BBL. Carolina Biological Supply has also proven to be a reliable source of cultures. The phage demonstration kits produced by Carolina Biological are highly recommended.

Although some programs may find it more convenient to use commercial cultures, quality control is, and always should be, the responsibility of the user.

USE OF PATHOGENS

There has been much discussion in recent years as to the advisability of using pathogenic or opportunistic organisms in the introductory microbiology laboratory. As much as it would be desirable to never use pathogens with first semester students, there are certain test reactions (i.e., coagulase) that cannot be suitably illustrated without the use of pathogens. For this reason, whereas their use has been minimized, they continue to be used when necessary.

It is the belief of the author that the most important lesson that can be left with all introductory microbiology students is that all organisms be treated as potential pathogens. With the careful emphasis on proper aseptic technique, the use of these organisms should not put students at risk.

MEDIA, STAINS, REAGENTS AND SUPPLIES

MEDIA

The following media are used in this manual. Most of them are available in dried as well as prepoured form. In most cases, the expense of the prepoured form is not justified, although convenience and quality may be overriding considerations in some circumstances.

Alkaline peptone medium
Bile-esculin agar*
Blood agar base* with 5% sheep blood
Brilliant green lactose bile broth*
Chocolate agar, with supplement*
Columbia CNA agar*
DNase agar, with methyl green*
EMB agar*
Hektoen enteric agar*
KF streptococcus broth*
Lauryl sulfate tryptose broth*
Levine EMB agar*
MacConkey's agar*
Mannitol salt agar*
m-Endo MF broth*
m-FC broth*
Minimal salts with glucose agar*
Motility medium*
MR-VP broth*
Mueller Hinton agar*
Mueller tellurite agar*
Nitrate broth*
Nutrient agar*
Nutrient broth*
Nutrient broth supplemented with 0.1% yeast extract and 1.0% glucose
Nutrient broth with pH 3.0, 5.0, 7.0, 9.0

Nutrient gelatin*
Peptone iron agar*
Phenol red broth base* with
    Glucose
    Lactose
    Mannitol
    Sucrose
Phenylethanol agar*
Plate count agar*
Rogosa SL agar*
Sabouraud agar*
Salmonella-Shigella (SS) agar*
Selenite F broth*
SIM agar*
Simmon's citrate agar*
Snyder test agar*
Starch agar*
Thayer-Martin agar*
Thioglycollate medium, fluid*
Thiosulfate citrate bile sucrose (TCBS) agar*
Triple sugar iron (TSI) agar*
Tryptone broth*
Tryptic soy agar (TSA)*
Tryptic soy agar + sodium chloride (5%, 10%, 15%)
Tryptic soy agar + sucrose (10%, 25%, 50%)
Tryptic soy broth*
Tryptic soy soft agar (0.7% agar)
Urea broth*

*Commercially available from Difco and BBL. For further information, refer to *Difco Manual of Dehydrated Culture Media and Reagents for Microbiology.* Difco Laboratories, Detroit, Ml 48232; and *BBL Manual of Products and Laboratory Procedures,* BBL, Becton Dickinson Microbiology Systems, Cockeysville, MD 2 1 030.

STAINS AND REAGENTS

Acetone/ethanol
Mix 100 ml of 95% ethyl alcohol and 100 ml of acetone.

Acid alcohol
Mix 3.0 ml of concentrated hydrochloric acid with 97.0ml of 95% ethyl alcohol

Albert's stain
Mix 0.2 gm of malachite green and 0.15 gm of toluidine blue in 2.0 ml of 95% ethyl alcohol. Add 100.0 ml of distilled water and 1.0 ml of glacial acetic acid.

Alpha-naphthol, 6%
See Barritt's Reagent A

Barritt's Reagent A
Dissolve 6.0 gm of alpha-naphthol into 100 ml of 95% ethyl alcohol.

Barritt's Reagent B
Dissolve 16 gm of potassium hydroxide in 100 ml of distilled water.

Coagulase plasma
Commercially available

Copper sulfate
Dissolve 20.0 gm of copper sulfate in 100.0 ml of distilled water.

Crystal violet
See Gram's Crystal Violet

Crystal violet, 1% w/v aqueous
Dissolve 1.0 gm of crystal violet in 100.0 ml of distilled water.

Dorner's nigrosine
Add 10.0 gm nigrosine to 100.0 ml distilled water. Mix together in a flask. Place the flask in a boiling water bath for 20 to 30 minutes. Add 0.5 ml formalin as a preservative. Filter twice through double filter paper.

Ferric chloride, 10%
Add 10.0 gm ferric chloride to 100.0 ml distilled water.

Gram's Crystal violet (Hucker's)
Stock solutions:

Solution A: Dissolve 2.0 gm of crystal violet in 20.0 ml of 95% ethyl alcohol.

Solution B: Dissolve 0.8 gm of ammonium oxalate in 80.0 ml of distilled water

Primary stain: Combine Solution A and Solution B. Allow to stand overnight. Filter before use.

Gram's iodine
Dissolve 1.0 gm of iodine and 2.0 gm of potassium iodide in 300 ml of distilled water. It may be necessary to grind the iodine and potassium iodide in a mortar while adding the water to the mixture.

Gram's Safranin
Dissolve 0.25 gm of Safranin 0 in 10 ml of 95% ethyl alcohol, then add the mixture to 100.0 ml of distilled water.

Gray's solution "A"
Stock solutions:

Solution 1: Dissolve 20.0 gm of tannic acid in 100 ml of distilled water.

Solution 2: Prepare a saturated solution of potassium alum.

Solution 3: Prepare a saturated solution of mercuric chloride.

Solution 4: Dissolve 3.0 gm of basic fuchsin in 100.0 ml of 95% ethyl alcohol.

Primary stain: Gray's solution "A" does not store well and should be prepared fresh daily. Combine the stock solutions in the following proportions:
No. 1 - 2.0 ml, No. 2 - 5.0 ml, No. 3 - 2.0 ml, and No. 4 - 0.4 ml. Filter before use.

Hydrogen peroxide, 3%

India ink

Use a good, drawing-quality India ink. The particles of carbon in the India ink should be as small as possible.

Iodine reagent-- Starch hydrolysis

Add 50 ml Gram's iodine to 50 ml distilled water.

Kinyoun's carbol fuchsin

Dissolve 4.0 gm basic fuchsin in 20.0 ml of 95% ethyl alcohol. Slowly add 100.0 ml distilled water while shaking. Add 8.0 ml phenol (heat to melt crystals).

Kovac's reagent

Combine 75 ml of n-amyl alcohol, 25 ml of concentrated hydrochloric acid, and 5.0 gm. of p-Dimethylamine benzaldehyde.

Loeffler's methylene blue

Dissolve 0.3 gm of methylene blue in 30.0 ml of 95% ethyl alcohol. Add 100.0 ml of distilled water.

Malachite green

Dissolve 5.0 gm of malachite green (oxalate) in 100.0 ml of distilled water. Filter before using.

Methyl red indicator

Dissolve 1.0 gm of methyl red in 300.0 ml of 95% ethyl alcohol. Add 500 ml of distilled water.

Methylene blue

Dissolve 0.3 gm of methylene blue in 100.0 ml of distilled water.

Methylcellulose

Mix 1.5 gm methyl cellulose in 100.0 ml distilled water.

Nitrate Reagent A

Dissolve 8.0 gm of sulfanilic acid in 1 liter of 5 N acetic acid.

Nitrate Reagent B

Dissolve 5.0 gm. of dimethyl-alpha-naphthylamine in 1 liter of 5 N acetic acid.

Oxidase Reagent

Dissolve 1.0 gm of dimethyl-p-phenylenediamine hydrochloride in 100.0 ml of distilled water. The reagent must be prepared fresh daily or stored frozen in 10.0 ml aliquots.

Potassium hydroxide, 20% with 5% alpha-naphthol

Combine 20% aqueous KOH containing 0.3% creatine and 3 drops of 5% alpha naphtha in ethyl alcohol.

Safranin

See Gram's safranin

Ziehl-Neelsen's Carbol Fuchsin

Stock Solutions:

Solution A: Dissolve 0.3 gm of basic fuchsin in 10.0 ml of 95% ethyl alcohol.

Solution B: Dissolve 5.0 gm of phenol crystals in 95.0 ml of distilled water.

Primary stain: Combine Solution A and Solution B.

Zinc dust

PREPARED REAGENTS, DISKS AND SLIDES

Prepared Reagents

Coagulase plasma

Reagent Disks

Bacitracin (Taxo "A")
Optochin (bile solubility)

Antibiotic Disks

Ampicillin, 10 mcg
Chloramphenicol, 30 mcg
Erythromycin, 15 mcg

Gentamycin, 10 mcg
Kanamycin, 30 mcg
Neomycin, 30 mcg
Novobiocin, 30 mcg
Penicillin, 10 units
Polymyxin B. 300 units
Streptomycin, 10 mcg
Tetracycline, 30 mcg

Antibiotic Solutions

Ampicillin
Chloramphenicol
Erythromycin
Gentamycin
Kanamycin
Penicillin
Polymyxin B
Streptomycin
Tetracycline

Serological Reagents

Blood Typing Sera
    anti-A
    anti-B
    anti-Rh

Serotyping Reagents
    Polyvalent *Salmonella* O Antiserum Set (e.g.,
    Difco Set: *Salmonella* Poly A- 1)

    Positive Control Antigen (e.g., Difco *Salmonella* O
    Antigen, Group B)

Febrile Agglutination Set with positive and negative
controls (e.g. Difco Set No. 2407-32-7)

Multiple Test Systems

API 20E (Analytab Products)
Enterotube II (Roche Diagnostics)

Miscellaneous Supplies

Disinfectant, laboratory
Ethyl alcohol, 95%
Mineral oil
Sheep blood
Xylol

Prepared Slides

Algae, mixed

Amoeba
Anabaena
Ascaris
Bacteria, 3 types mixed
Blood smear, stained
Capsule stain
Clonorchis
Corynebacterium diphtheriae, metachromatic
granule stain
Enterobius
Escherichia coli, Gram stain
Euglena
Flagella stain
Giardia
Gleocapsa
Letter 'e'
Mycobacterium tuberculosis, acid fast stain
Negative stain
Paramecium
Plasmodium
Pneumocystis carinii
Schistosoma
Spore stain
Staphylococcus aureus, Gram stain
Taenia
Trichinella
Trypanosoma

Preserved specimen

Ascaris
Clonorchis
Enterobius
Schistosoma
Taenia
Trichinella

SUPPLIES

Anaerobic Culturette© Collection and Transport
    System (Becton Dickinson and Co)
Antibiotic disk dispenser
Artificial blood
Bacturcult Culture Tube (Wampole Laboratories)
Breed slide
Calcium alginate swabs
Calibrated loop
Candle jar
Counting chamber: Neubauer or Petroff-Hauser
Coverslips
Culturette© Collection and Transport System (Becton

Dickinson and Co)
Gas-Pak anaerobic jar with catalyst, indicator, and
    gas-generator envelopes
Glass spreader
Kimwipes
Lancets
Latex gloves
Membrane filter set-up
Membrane filters, 0.45 µm pore size
Microscope slides
Ocular micrometer

Paper disks, 6.0 mm dia.
Paraffin block
Pasteur pipettes
Pipetter
Pipettes: serological, 1.0 ml, 5.0 ml, 10.0 ml
Replica plate
Slide micrometer
Spectrophotometer and cuvettes
Stainless steel pins
Swabs
Ultraviolet light

APPENDIX C
REFERENCES

RESOURCE REFERENCES

The following resource references provide a more detailed, in-depth explanation of the various procedures or further discussion of the characteristics of the various organisms used.

1. Baron, E.J., L.R. Peterson and S.M. Finegold. *Bailey and Scott's Diagnostic Microbiology,* 9th Ed. St. Louis MO: C.V. Mosby Co., 1994.

2. Gerhardt, P., editor-in-chief. *Methods for General and Molecular Bacteriology.* Washington DC: American Society for Microbiology, 1994.

3. Holt, J.G., editor-in-chief. *Bergey's Manual of Determinative Bacteriology,* 9th Ed. Baltimore: Williams and Wilkins Co., 1994.

4. Holt, J.G., editor-in-chief. *Bergey's Manual of Systematic Bacteriology.* Baltimore: Williams and Wilkins Co., 1984.

5. Isenberg, H.D., editor-in-chief. *Clinical Microbiology Procedures Handbook.* Washington DC: American Society for Microbiology, 1992.

6. Koneman, E.W. et al. *Color Atlas and Textbook of Diagnostic Microbiology,* 4th Ed. Philadelphia: JB Lippincott Co., 1992.

7. Murray, P.R. et al, editors. *Manual of Clinical Microbiology,* 6th Ed. Washington DC: American Society for Microbiology, 1995.

TEXT REFERENCES

The following texts are referenced as to the appropriate sections of your text. See the following table for the correlations to specific exercises.

1. Atlas, R.M. *Microorganisms in our World.* St. Louis MO: Mosby, 1995.

2. Black, J.G. *Microbiology: Principles & Applications,* 3rd Ed. Upper Saddle River NJ: Prentice-Hall, 1996.

3. Brock, T. D., M .T. Madigan, J .M. Martinko and J .Parker. *Biology of Microorganisms,* 7th Ed. Englewood Cliffs NJ:Prentice-Hall, 1994.

4. Ingraham, J.L. and C.A. Ingraham. *Introduction to Microbiology.* Belmont CA: Wadsworth Publishing Company, 1995.

5. McKane, L. and J. Kandel. *Microbiology: Essentials and Applications,* 2nd Ed. New York NY: McGraw-Hill, Inc., 1996.

6. Nester, E.W., C.E. Roberts and M.T. Nester. *Microbiology: A Human Perspective.* Dubuque IA: Wm. C. Brown Publishers, 1995.

7. Talaro, K. and A. Talaro. Foundations in Microbiology, 2nd Ed. Dubuque IA: Wm. C. Brown Publishers, 1996.

8. Tortora, G.J., B.R. Funke and C.L. Case. *Microbiology: An Introduction,* 5th Ed. Redwood City CA: The Benjamin/Cummings Publishing Co., Inc.,1995.

EXERCISE	ATLAS	BLACK	BROCK	INGRAH.	MCKANE	NESTER	TALARO	TORTORA
1. Intro. to Microscopy	3	3	3	3	3	3	3	3
2. Prep. of Smears, Simple Stains, and Wet Mounts	3	3	3	3	3	3	3	3
3. Differential Staining	3, 5	3, 4	3	3, 4	3, 4	3	3, 4	3, 4
4. Additional Staining	3, 5	3, 4	3	3, 4	3, 4	3	3, 4	3, 4
5. Calibration of Micro.	3	3, A	3	3	3	3	3	3
6. Cell Dimensions	3	4	3	3	4	3	4	4
7. Survey of Microorganisms	2, 23	4, 12	21	12	2, 11, 12	12	5	4, 12
8. Isolation Techniques	3	6	4, 17	3	3	4	3	6
9. Oxygen Requirements	10	6	9	8	5	4	7	6
10. Physical Growth Requirements	10	6	9		5	4	7	6
11. Selective & Differential Media	3	6	13		5	4	3	6
12. Carbohydrate Metabolism	6	5	4, 16	5	6, 7	5	8	5
13. Nitrogen Metabolism	6	5	4, 16	5	6, 7	5	8	5
14. Misc. Rx	6	5	4, 16	5	6, 7	5	8	5
15. Development of Diagnostic Key	2	9, 10, B	18	10, 11	10	10, 11, B	1, 4	10, 11, A
16. Use of Diagnostic Key	2	9, 10, B	18	10, 11	10	10, B	1, 4	10, 11, A
17. Rapid Diagnostic Tests	16	6	13					6
18. Microscopic Counting Methods	10	6	9	8	5	4	7	6
19. Viable Cell Counts	10	6	9	8	5	4	7	6
20. Spectrophotometric Methods	10	6	9	8	5	4	7	6

EXERCISE	ATLAS	BLACK	BROCK	INGRAH.	MCKANE	NESTER	TALARO	TORTORA
21. Physical Control Methods	11	13	9	9	15	9	11, C	7
22. UV Light as Antimicrobial	11	13	9	9	15	9	11	7
23. Assay of Antimicrobials: Disk-Diffusion Method	11	13	9	9	15	9	11, C	7
24. Assay of Antimicrobials: Use-Dilution Method	11, 17	13	9	9	15	9	12, C	7
25. Antibiotic Sensitivity Testing	11, 17	14	9, 13	21	16	29	12	20
26. Detection of Mutant Strains	7	7	7	6	8	7	9	8
27. Epidemiology	12	16	14	20	20	18	13	14
28. Serological Reactions	16	19	13	19	18	17	16	18
29. Collection and Transport of Specimen	16	D	13		C		Intro. to Med. Micro.	C
30. Throat Cultures	16, 22	21	11	22	22	22	18, F	24
31. Urinary Tract Cultures	16, 22	25	11	24	23	24	18, 20, F	26
32. Gastrointestinal Tract Cultures	16, 20	22	11	23	22	23	20, F	25
33. Lactobacillus Activity in Saliva	22	22	11	23	22	22	21	25
34. Viral Population Counts	9	11	6	13	13	13	6	13
35. Food Microbiology	24	27	17	9	27	33	26	28
36. Water Analysis	25	26	17	28	26	31, 32	26	27

Figure 1. Hanging-drop preparation.

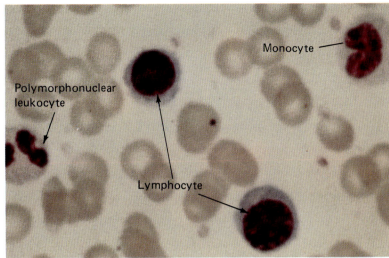

Figure 2. Human blood smear.

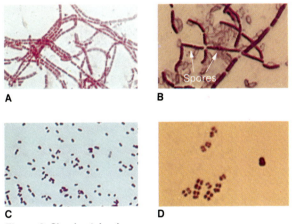

A
B
C
D

Figure 3. Simple stained smears.

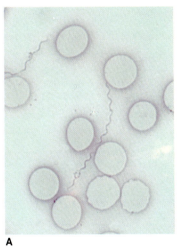

A
Figure 4A *Borrelia recurrentis.*

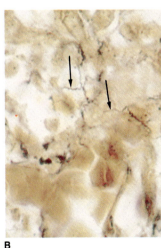

B
Figure 4B. *Treponema pallidum.*

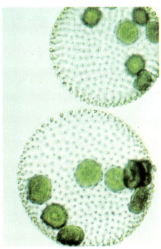

Figure 5. *Volvox.*

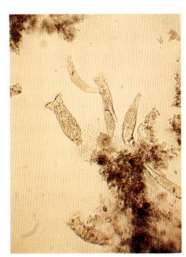

Figure 6. Rotifers.

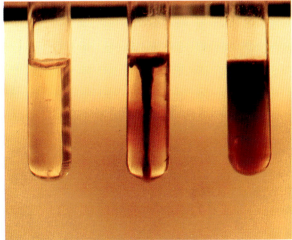

Figure 7. Motility medium.

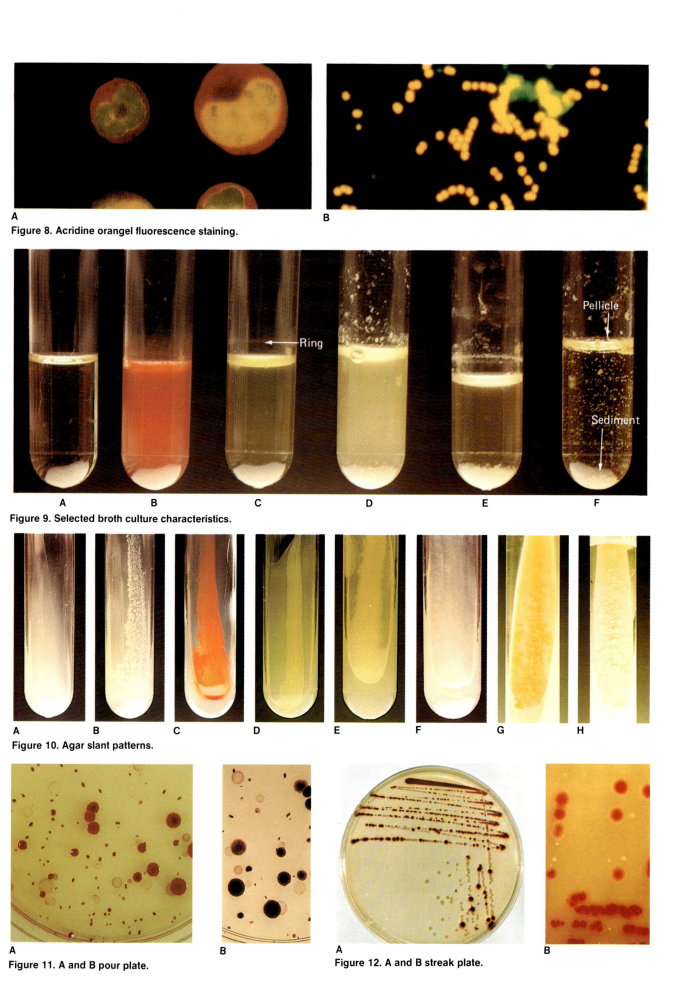

A

B

Figure 8. Acridine orangel fluorescence staining.

A   B   C   Ring   D   E   Pellicle

Sediment

A   B   C   D   E   F

Figure 9. Selected broth culture characteristics.

A   B   C   D   E   F   G   H

Figure 10. Agar slant patterns.

A   B

Figure 11. A and B pour plate.

A   B

Figure 12. A and B streak plate.

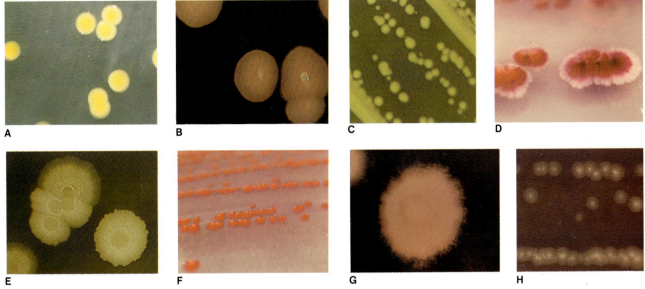

A

B

C

D

E

F

G

H

Figure 13. Bacterial colony characteristics.

Figure 14. Differentiation of *Pseudomonas.*

Figure 15. *Pseudomonas.*

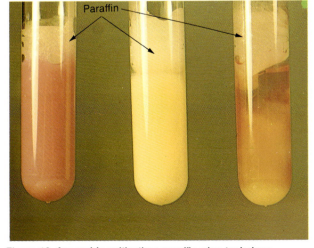

Paraffin

Figure 16. Anaerobic cultivation: paraffin-plug technique.

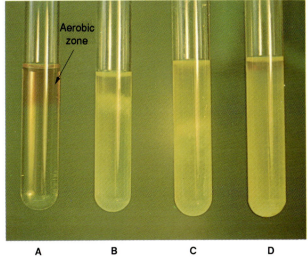

Aerobic
zone

A   B   C   D

Figure 17. Fluid thiglycollate broth lpatterns.

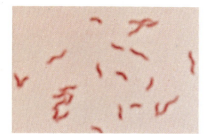

Figure 18. *Vibrio cholerae.*

A

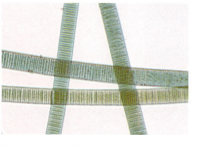

Figure 19. *Oscillatoria.*

B

Figure 20. A and B.
Representative algae.

Figure 21. *Alternaria.*

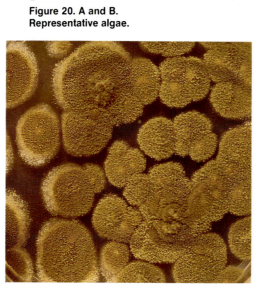

Figure 22. *Aspergillus niger.*

Figure 23. *A flavus.*

Figure 24. *Fusarium.*

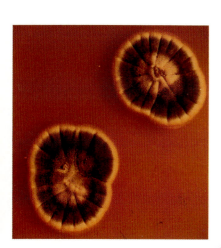

Figure 25. *Penicillium notatum.*

Figure 26. *Rhizopus nigricans.*

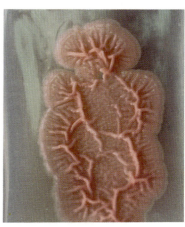

Figure 27. *Rhodoturula.*

Figure 28. *Saccharomyces cerevisiae.*

Figure 29. *Scopulariopsis.*

Figure 30. Mushroom structure.

Figure 31. Spore print.

Figure 32. Lichens.

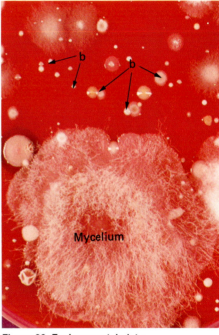

Figure 33. Environmental plate.

Figure 34. Environmental sampling flask.

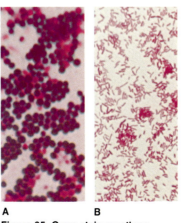

A   B
Figure 35. Gram stain reactions.

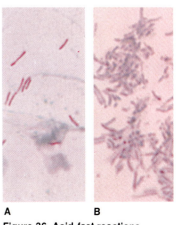

A   B
Figure 36. Acid-fast reactions.

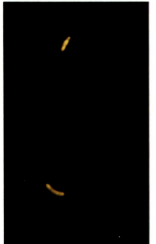

Figure 37. FA-F.

Figure 38. Capsules.

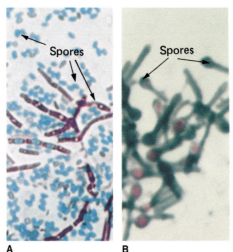

A

Figure 39. Spores.

B

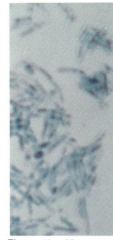

Figure 40. Meta-chromatic granules.

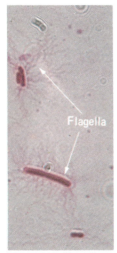

Flagella

Figure 41. Flagella.

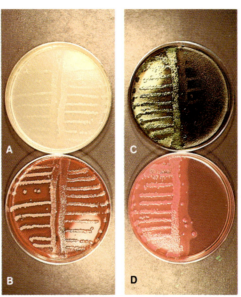

Figure 42 Differential and selective media.

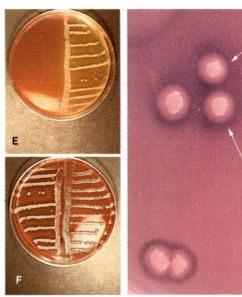

Figure 43 Caseinase activity.

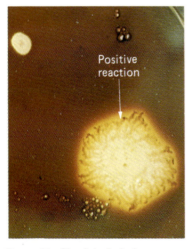

Positive reaction

Figure 44 Starch hydrolysis.

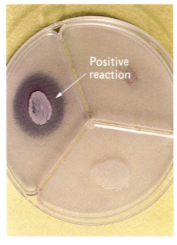

Positive reaction

Figure 45 Lipid hydrolysis.

Gas

A   B   C

Figure 46 Tube carbohydrate fermentation reactions.

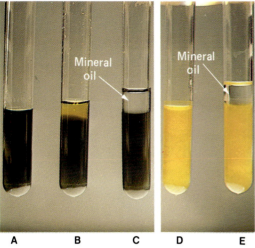

A   B   C   D   E

Figure 47   Carbohydrate oxidation-fermentation reactions.

Figure 48A.   Gelatin hydrolysis.

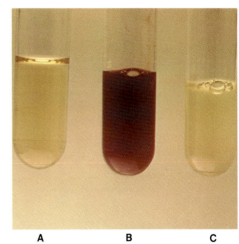

A   B   C

Figure 48B.   Nitrate reduction.

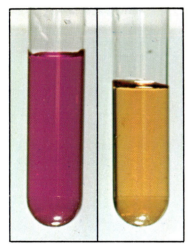

Figure 48C.   Urease production.

Figure 49.   Catalase reactions.

A   Indol test
Figure 50.   IMViC reactions.

B   Methy red test
Figure 50. IMViC reactions.

C   VP test

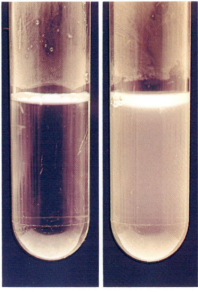

D   Citrate test

E   Citrate test

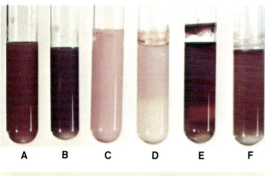

G

Figure 51. Selected litmus milk reactions.

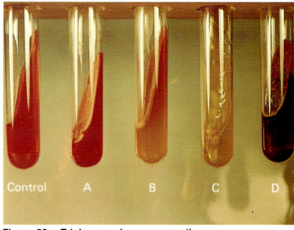

Figure 52. Triple sugar iron agar reactions.

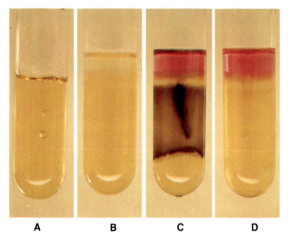

A   B   C   D

Figure 53. Bacto-SIM reactions.

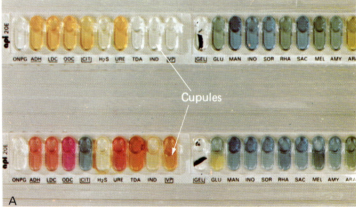

Figure 54. Multiple test systems.

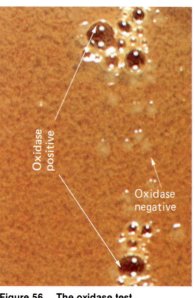

Figure 55. PA reactions.

Figure 56. The oxidase test.

Figure 57. UV lethal effects.

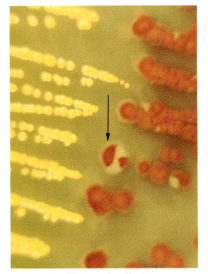

Figure 58.   Mutagenic effects.

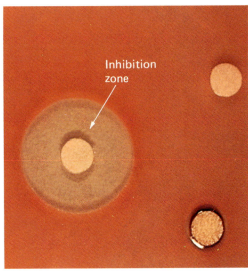

Figure 59.   Oligodynamic action.

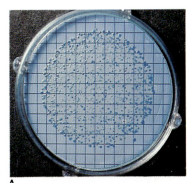

A

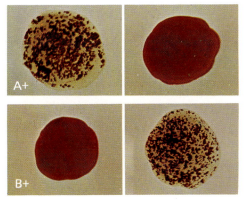

B   Figure 60.   Membrane filter technique.

Figure 61. Bacteriostatic activity of dyes.

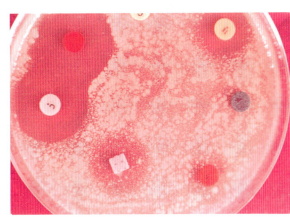

Figure 62. Antibiotic sensitivity testing.

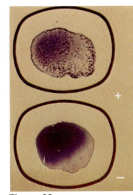

Figure 63. Coagglutination.

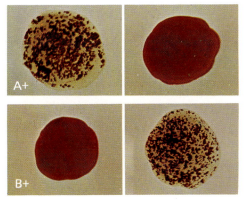

Figure 64. Typical blood agglutination reactions.

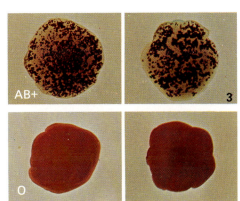

Figure 65. Rh_0 reactions.

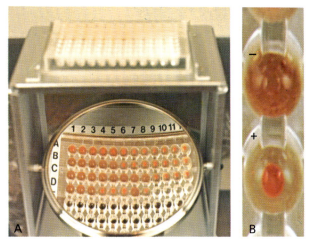

A

B

Figure 66. Hemagglutination inhibition reactions.

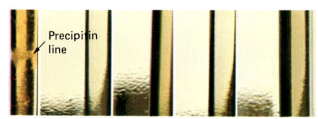

Figure 67. The precipitin reaction.

Figure 68. Complement fixation reactions.

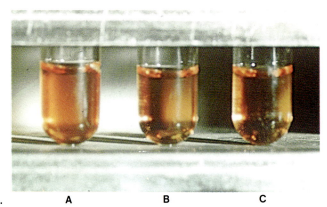

A   B   C

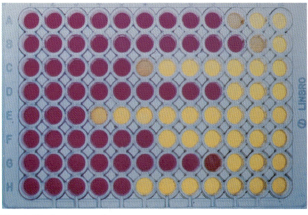

Figure 69. An enzyme immunoassay.

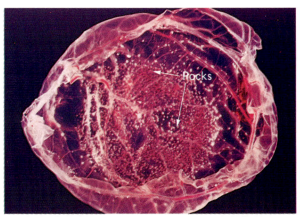

Figure 70. Pock formation on chorioallantoic membrane.

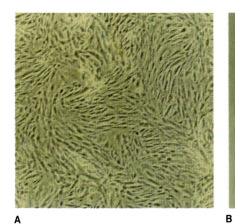

A   B

Figure 71. Virus infected tissue cultures.

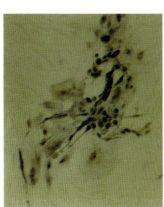

Figure 72. Gall formation.

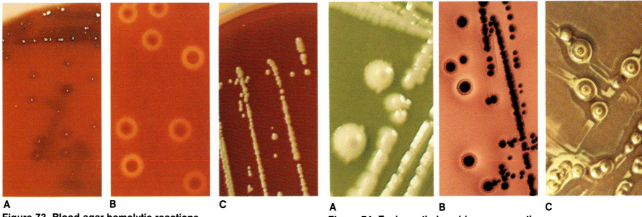

A   B   C   A   B   C

Figure 73. Blood agar hemolytic reactions.   **Figure 74. Eosin-methylene blue agar reactions.**

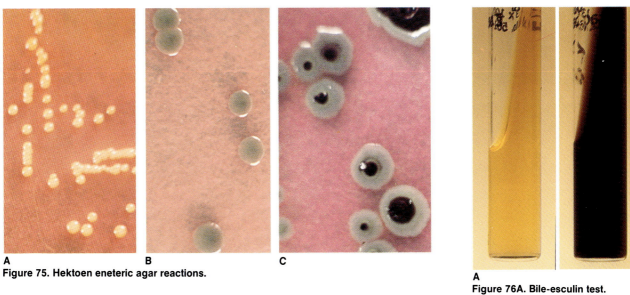

A   B   C

Figure 75. Hektoen eneteric agar reactions.

A

Figure 76A. Bile-esculin test.

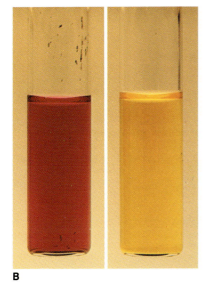

B

Figure 76B. Salt tolerance

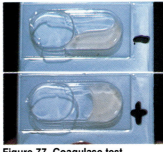

Figure 77. Coagulase test.

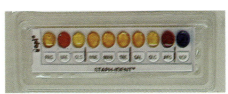

Figure 78. Staph-Ident strip.

Figure 79. Staph latex test.

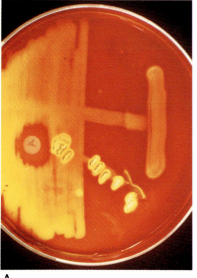  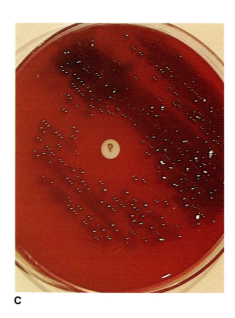

A   B   C

Figure 80. Bacitracin sensitivity, CAMP, and Optochin tests

  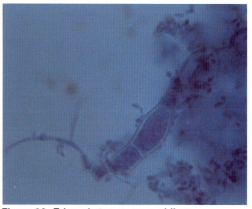

Figure 81. Mycellim of
Microsporum canis.

Figure 82. Mycellim of
Epidermophyton floccosum.

Figure 83. ***Tricrophyton macroconidium.***

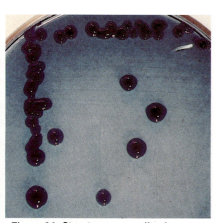

Figure 85. ***Streptococcus mitis*** **on mitis-salivarius medium.**

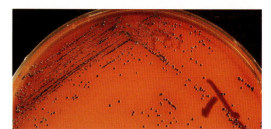

Figure 84. ***Streptococcus salivarius***
mitis-salivarius medium.

Figure 86. ***Corynebacterium xerosis*** **on**
Mueller-Hinton-tellurite medium.

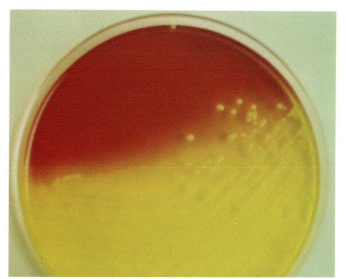

Figure 87. Pathogenic *Staphylococcus aureus* on mannitol-salt agar.

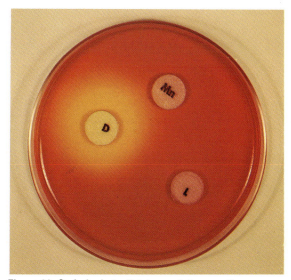

Figure 88. Carbohydrate disk fermentation reactions.

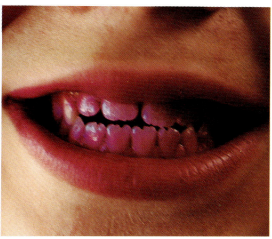

Figure 89. Dental plaque detection.

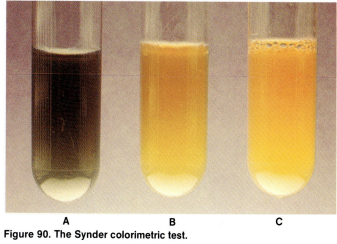

A   B   C

Figure 90. The Synder colorimetric test.

Figure 91. Mycelium of *Coccidioides immitis.*

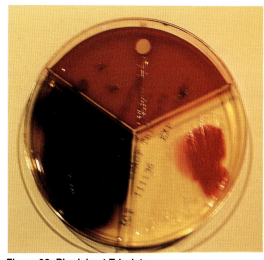

Figure 92. Phadabact Tri-plate.

Figure 93. Beta-hemolysis by agar stab.

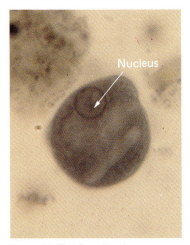

Figure 94. Trophozoite of *Entamoeba histolytica.*

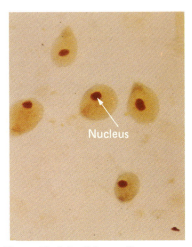

Figure 95. *Balantidium coli.*

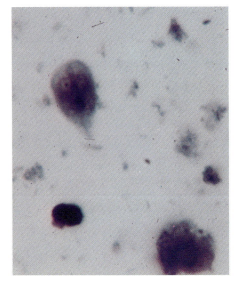

Figure 96. *Giardia intestinalis.*

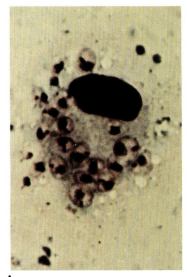

A
Figure 97. Leishmania tropica.

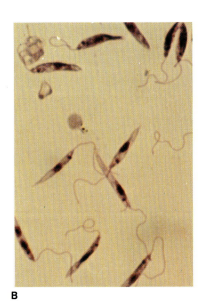

B

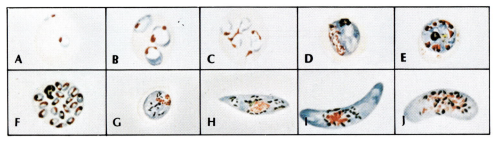

Figure 98. *Plasmodium falciparum.*

Figure 99. *Plasmodium malariae.*

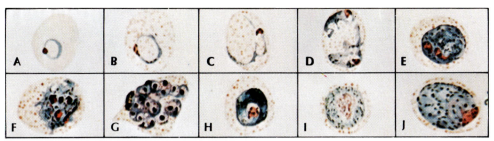
Figure 100. *Plasmodium vivax.*

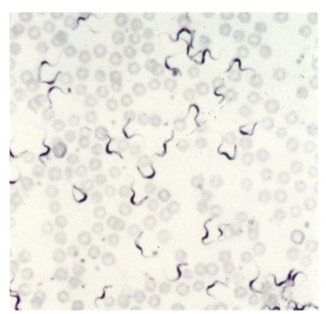
Figure 101A. *Trypanosoma brucei gambiense.*

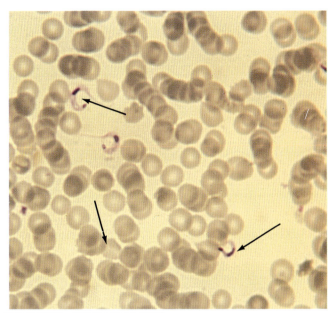
Figure 101B. *Trypanosoma cruxi.*

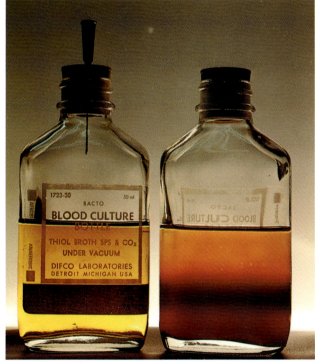
A   B
Figure 102. Blood culture bottles.

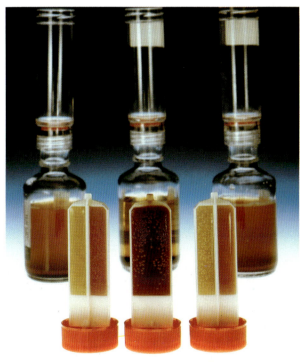
Figure 103. Septi-chek system.

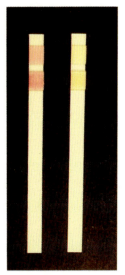

Figure 104. Chemstrip
LN reactions.

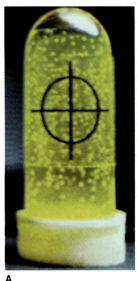

A

B

C

Figure 105. Bacturcult reactions.

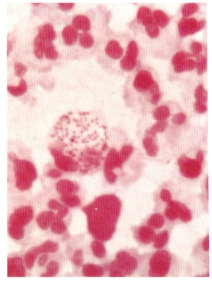

Figure 106. *Neisseria gonorrhoeae.*

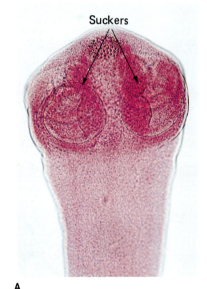

Suckers

A

Hooklets

Proglottid

B

Figure 107. Tapeworms.

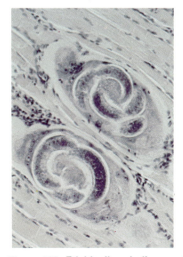

Figure 108. *Trichinella spiralis.*

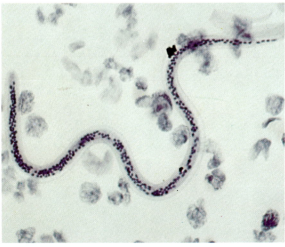

Figure 109. *Wuchereria bancrofti.*

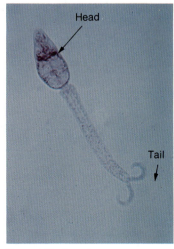

Head

Tail

Figure 110. Ceraria.